Healing Tips

for the Mind, Body, and Soul

Arif Shaikh R.Ph. M.D.

Healing Tips
for the Mind, Body, and Soul

Copyright © 2023 by Arif Shaikh R.Ph. M.D.

Paperback ISBN: 978-1-63812-566-2
Ebook ISBN: 978-1-63812-567-9

All rights reserved. No part in this book may be produced and transmitted in any form or by any means, electronic, or mechanical, including photocopying, recording, or by any information storage and retrieval system, without permission in writing from the copyright owner.

The views expressed in this work are solely those of the author and do not necessarily reflect the views of the publisher hereby disclaims any responsibility for them.

Published by Pen Culture Solutions   01/30/2023

Pen Culture Solutions
1-888-727-7204 (USA)
1-800-950-458 (Australia)
support@penculturesolutions.com

Letters to the Author

"You are my friend as well as my physician. The book *Healing Tips* is a plethora of information. Many talks we've shared over the years have brought to light and many pieces of advice for both of us. As Ben Franklin once said, 'Something well done is better than well said.'"
—*John Cavalier*

"I loved your book *Healing Tips*. I have been meaning to tell you how beneficial my visits to you have been. Your book and your encouragement and overall positive approach have been profoundly helpful in restoring my self-image and providing insights into the root cause of my problem."
—*Stan Stohl*

"Thank you for your insightful book. It has been very helpful to me as I reflect on choices I have made in my life and the consequences and outcomes that have followed. The healing tips in your book, on your website and when I see you in the office make me realize some of the destructive behaviors I have allowed in my life and MY ability to make positive changes. It is no one else's fault but mine if I can't be happy. No one controls my happiness except me—I am the only person that can allow someone to take my happiness away."
—*A. F.*

"Dr.Shaikh,

Your book Healing Tips has been extremely helpful and insightful to me. I'ts a great guide for difficult times for people everywhere. Certainly it has helped me a lot with my circumstances. Many points taught me that foremost we are all important, be grateful, find purpose through suffering, take care of the moment, judge state of mind control what you see, develop immunity to offenses, condemntion and puts downs. Dont judge successes by outcome, learn from others, and worship God through service, hard work and love. Think positive thoughts, stay strong and focus on the goal.

You provide inspiration and guidance to so many of us.When so much turns against us, God sends you full og light and wisdom. Challenge after challenge has visitedus. Your words come through the clouds. You provide faith, hope, and love. It comes on a plate of humility. Your words turns trash into beauty. You gave reason to go on, the way to do it, and the genuineness to help everybody listen. Your words are remembered; your life is admired.

Thanks!"

—Jerry Sandusky

Note from the Editor

How I see the world:

From where I sit and view the world, It helps me constantly change.

I become a physician because I wanted to help people, only to realize, that I cannot help anyone; but learn from everyone.

I became a physician because I wanted to help people. As I pursued my quest, I was humbled to realize that I was learning from people much more than I was helping them.

I find this to be true.

I am reminded of Jesus saying, "Physicians thy heal themselves."

Patients are just mirrors, showing me my sick self, and helping me grow personally.

I am grateful to each one for helping me be who I am. I love that.

Healing tips are basically what I learned from others and therefor like to share, hoping to have their input, so I may keep learning.

The letters I received are lot more powerful than the book I wrote.

This book is just a talk, but the letters from others are the walk, and the walk is much louder!

I thank my Creator, who helps me all along, especially my parents, my siblings, my wife, my children, co-workers, family, friends, my patient, and all humans in general.

Healing Tips for the Mind, Body, and Soul

Healing Tip 1

Life is a honeymoon for you and your body. If you do not take care of your body, your honeymoon is over!

P.S. Therefore, take care of your body and enjoy your honeymoon!

Healing Tip 2

Treat everyday like it's a vacation, no matter where you are. Keep your friends close to you throughout these special days, and take good care of them.

Otherwise, your vacation may end.

Healing Tip 3

Be captive of your own walls and boundaries.
If necessary, break the walls and redefine your boundaries!
By doing so I am now the king or Queen of my Kingdom & Live happily ever after!

Healing Tip 4

You were born with this body & will be together until death separates you.

If you love and take care of your body, then you will be on a honeymoon for the rest of your life.

Also remember that your body loves you! Follow your commands and it never sends you a bill!

Healing Tip 5

Normal aging will make you old.

Living a healthy life gracefully and you will earn a "G", old age will be golden years.

Healing Tip 6

Newborns have a future because they have no past.
Adults with sinful pasts have no future unless they repent.
And do self-purification.

Healing Tip 7

Every newborn is a MASTERPIECE.
Every adult is a GOLD MINE.
If you dig into your past and learn from your mistakes and refine yourself—you will turn into pure gold.

Healing Tip 8

Sometimes it is okay to be selfish. If you give everything for free you will end up being bankrupt.
And become reasonable person

Healing Tip 9

Self-awareness will bring peace. Become self aware and become a peacemaker.
Faulty self-awareness will cause judgmental attitudes towards others, and this can cause trouble.
Unawareness of thy self will create anxiety, stress, depression and turmoil.
This way of life will not bring peace.

Healing Tip 10

Helping one human being including yourself is like helping all of humanity.
Hurting one human being including yourself is like hurting all of humanity.

Healing Tip 11

Be Grateful!
Being grateful for everything you have. This will bring you happiness, peace and success.
Being ungrateful is a sign of not being self aware, and this causes unhappiness.

Healing Tip 12

Love and Hate...
Love or Hate attaches you to anything or anyone.
These attachments can affect the quality of your life.
You should pay attention to the underlying reasons for your attachments.

Healing Tip 13

Purpose...
General expectations gave me suffering.
Suffering was the result of assumed expectations.
Through suffering I found purpose.
Your life will suffer if you find no purpose.

Healing Tip 14

Present Moment...
The past is history.
The future is a mystery.
The present moment holds the key to make history, and to discover the future mystery.

Healing Tip 15

Health Account Balance...
Doing healthy things is like making deposits to a health account.
Doing unhealthy things is like making withdrawals from a health account.
Your health account balance determines the quality and quantity of my life.

Healing Tip 16

Gifted and Blessed...
We are all in this world together.
This world is full of blessed and gifted people.
Blessed are the poor, needy, sick and prisoners.
Gifted people can earn their blessings by serving blessed people, and by helping them to fulfill their needs, therefore to earn their gift of health, wealth and freedom.
If gifted people do not use their gifts wisely, they will end up losing their gifts, and the blessed people will be deprived.

Healing Tip 17

You are gifted.
These gifts are like time bombs.
If you use these gifts in a timely fashion, you will receive blessings by serving blessed people to earn the gifts they need.
Otherwise their gifts will explode, and you will be blasted away, leaving the rest deprived.
Do you want to be blessed and gifted?
OR
Do I want to be blasted and grieved?

Healing Tip 18

Perfect Universe...
You are perfectly imperfect.
This universe is perfect.
If you see this universe with an imperfect lens, it will look imperfect.
If you become aware and see the universe the way it is, then
You will be able to see perfection in everything and everyone.

Healing Tip 19

If you save one life and heal one soul during my lifetime —
You have served your purpose.

Healing Tip 20

Every human being has unique shortcomings and unique strengths. If I become self-aware, I will know my shortcomings, and will be able to see others who do not have those shortcomings. I will learn from them.

At the present I am not doing okay and others are okay. Maybe one day I will be okay.

If you are not doing okay, seek help. There are always other who are okay, and a day will come when you are better.

Healing Tip 21

(Self Love + Discipline + Patience) (Diet + Exercise + Good sleep) (Health) = HAPPINESS

(Unpaid worry + Procrastination + Self-hatred) (Unawareness) = unhappiness

happiness − unhappiness = general wellbeing

(Quality of happiness depends upon the purity or impurity of the ingredients)

INDICATIONS: Found to be very effective against multiple mental, emotional and physical ailments.

SIDE EFFECTS: None yet reported.

CONCLUSION: HAPPINESS IS A GOOD MEDICINE!

Healing Tip 22

I must love my enemy, so that I do not have an enemy.
You are your own worst enemy!

Healing Tip 23

Baby talk sounds great, but does not make much sense.

Some grown up talk sometimes does not make much sense either. A grown ups walk is always louder and clearer than their talk.

P.S. If someone has trouble hearing a walk, then a kick in the butt may restore the hearing.

Healing Tip 24

You are run by the congress: 50% Angel party, 50% Demon party.

You have the deciding vote.

You can be an Angel one moment, and a Demon the other.

You are the king or queen of my kingdom.

Healing Tip 25

(QUALITY OF MY AWARENESS) +

(QUALITY OF MY HAPPINESS) (QUANTITY OF MY SHORTCOMINGS)

Without awareness, success, money, power, health, beauty or material things may make me comfortably unhappy.

Healing Tip 26

Computers have hardware and software; both must be compatible.

Humans have a body and a mind; both must be compatible.

For best results, hardware needs recommended maintenance, and software needs recommended upgrading.

The same is true for human body and mind.

Well-maintained computers can interface with other computers.

Well-maintained humans can relate with other humans.

If you follow the recommendations 100%: your mind, body, and computer will work like a team, and you will be 100% happy with your mind, body and computer.

Healing Tip 27

The Soul...
If you cannot see clearly with your eyes closed;
Either your heart is blind, or you need a new light bulb.

Healing Tip 28

Grow with the World...
The world will be to me as I am to the world.

If you are helping to make this world better, this world will help you grow, and you will have peace and security.

If I am just looking out for myself and do evil in this world, then I will be in a state of war and live in fear.

Healing Tip 29

Self Growth...

Ignorance and lack of self-awareness cause mostly self-destruction.

Knowledge and self-awareness cause mostly self-growth.

The more you know about yourself, the more knowledgeable and wiser you shall become.

Healing Tip 30

Quit Smoking...

Life story of a smoker

I loved puffing, I started huffing, and I puffed and huffed.

I was able to do less puffing and more huffing.

Now I cannot puff or huff.

My advice to everyone: Please do not start puffing. If you already puff, please stop.

From EX Smoker: Graveyard St., Grave #420, Harrisburg, PA 17110

1-800-Smokers - Ext- Loser

www.SmokersDieYoung.com

Healing Tip 31

I feel so unappreciated and neglected.
Please do not make me feel this way.
Visit me and talk to me as this may be a start of an intimate relationship that may last a lifetime.
I am only yours, serving you unconditionally and your wish is my command.
Love,
YOUR BODY

Healing Tip 32

Be Happy, Successful and Peaceful...
You were born to be happy, successful and peaceful.
You are the only one who can decide otherwise.
 Please visit your sick soul before it is too late!

Healing Tip 33

Learning from experience...
Somehow I always felt that the experience I gained was not needed for me.
I always shared my mistakes with others, so that they would not make the same mistakes.
In fact, I was eager to teach others, so they can learn from my mistakes and experiences.
No one seemed interested in learning from me. I felt sorry for them.

Later in life I started feeling sorry for myself. I too should have learned not only from my mistakes, but from other people and should have practiced what I learned.

You should learn from my mistakes and from other people and practice what you learn in order to live a successful, happy and peaceful life.

Healing Tip 34

Recipe to Fix a Broken Heart...

A long time ago, Jack stole innocent Jackie's heart. He tore her heart apart and left it shattered.

Heartbroken, Jackie did not know what to do.

She researched and worked very hard and was able to discover the recipe to fix a broken heart.

It was hard, but Jackie got rid of all the defective parts of the heart and was able to make a new and improved healthy heart.

She was a happy girl.

Jack went out with Jill, fell down the hill and broke his neck.

Jack started cursing at Jill. Jill got angry called him "Jackass" and left.

With no place to go, Jack went back to Jackie.

Jackie took good care of Jack, and he got physically healthy.

Jack stole Jackie's heart and tried to break it apart, but could not because it was a strong healthy heart.

Jack ended up giving his heart to Jackie. With Jackie's help they were able to rebuild Jack's heart without defective parts.

Now Jack & Jackie were two bodies and one heart and lived happily ever after.

Healing Tip 35

Recipe to Live Without Worry...
If you always keep up, no worries ever catch up.

Healing Tip 36

If you see yourself in the mirror, you are only seeing the tip of the iceberg.

If you see an unhealthy body, sick mind, and an agonizing soul, this may be the whole iceberg.

Awareness is the main ingredient of any successful treatment plan for body, mind and soul.

Healing Tip 37

Your life is not perfect, neither is your health.

Healing Tip 38

With perfect health I can move mountains.

To acquire perfect health, you have to move myself to a healthy state of mind.

Sometimes this is harder than moving mountains.

Healing Tip 39

For Relationships...
Behind most successful men there is a woman.
Behind most unsuccessful men there is more than one woman.
Behind most successful women there is a successful and happy family.
Behind most unsuccessful women there is a man.
MORAL: I must guard my behind and be very kind to the special one who is guarding my behind.
Without a strong behind, I cannot have strong enough legs to stand on.

Healing Tip 40

For Self-Guidance...
With self-awareness you will discover the GPS in you.
To activate and put a destination in, you have to be in a healthy frame of mind.
If you are, life will be a pleasant journey, smooth ride, and you will reach my destination safely.

Healing Tip 41

For Self-Awareness…

You were born great.

If you would have stayed in perfect surroundings (which is and was impossible), then you would be a perfect human being.

Become self-aware and learn the things that make you imperfect were not necessarily bad; it was a boot camp to make you who you were meant to be.

Healing Tip 42

For Keeping Yourself Together…

If I am healthy, I will keep my mind in a good frame and my body in good shape.

I have to stay one step ahead so I can be a great guide.

If you do not guide your mind, get out of the frame, do crazy things, have your body out of shape, and do wild things- you will be in trouble.

If you move too fast, you may leave your mind and body behind wondering. They may go their own way. You may get lost, and you will have to go back to find them.

Healing Tip 43

On honesty…

If you become brutally honest with yourself, then you will discover many things that you never wanted to discover.

Healing Tip 44

For Yourself...

You will know what you have to do, what you do not want to do, or do not have enough courage to do.

If you do not have enough drive to overcome unreasonable desires, then you will never grow, and you will stay in an unhappy uncomfortable state.

Healing Tip 45

Self Discovery...

If you acquire the ability to block your five senses and isolate your mind so no outside impulse can penetrate or distract, you will find a universe ready to be explored and discovered.

That will be the most rewarding experience you have ever had.

Healing Tip 46

True success cannot be gauged by outcome.

If person A can crawl one mile an hour, and he or she is doing one mile an hour, person A is doing 100%.

If person B can run 10 miles an hour, and he or she is doing 6 miles an hour, person B seems six times faster than person A, but is doing 60%.

Getting to know your capacity is the first step towards success.

Healing Tip 47

Your life is meant to be purposeful, happy, healthy, and successful.
You best friends are always with you: your mind and body.
You must detoxify your mind, and heal your body.

Healing Tip 48

Life is a journey, and awareness is the radar.
Every event in life that is pleasant can be a distraction, and if unpleasant can be a reminder on the awareness screen.
If you watch the radar of your awareness, then you will reach your destination safely.

Healing Tip 49

Life is meant to be a vacation.
My two best friends are always with me: my mind and body. I must take care of my friends; otherwise, I may lose my mind, and my body will give up on me.
My vacation will be over.

Healing Tip 50

Finding a balance is the key to success.
I, my, me, and myself.
I and me must watch myself.
I may have to keep both eyes on myself to stay away from trouble.
I have a choice to stay in a self-denial state and be controlled by "My", but if I can tame "my" part of myself, that is self-purification.

Healing Tip 51

(BEWARE OF HUMAN)
I know I am living with myself, and this is not easy.
"My" part of myself is out of control, and my poor "self" is weak and dirty.
I have to work hard for SELF purification; then there will be a balance between "MY" and "SELF."
I will live a happy, balanced, and successful life.

Healing Tip 52

A correct diagnosis is the key to a successful treatment plan.

If you know your correct diagnosis, it will be the start of a happy and successful life.

If you do not, you an call human mechanics at 1-800-PHY-SICIAN.

If you do not correct yourself or do not contact the human mechanics in a timely manner, you may be recalled before my time.

Healing Tip 53

I must make a conscious effort to do well, to see well.
I must make a conscious effort to do no bad, to see no bad.
I will become healthier and stay healthy.
If I am unaware, I will do bad, see bad.

Healing Tip 54

Good health + Sound Mind = Perfect life
Good health = Diet + Exercise + Sleep = Productive life
Sound mind = Diet (good healthy thoughts) + Healthy mental exercises + Restful sleep = Sweet dreams.

Healing Tip 55

I used to drive myself crazy.
My license got revoked and suspended.
I did my soul searching and walked myself to a peace garden.
I came to terms with myself.
Now I have a restrictive license and am only allowed to drive myself to heaven.
You have to power to choose which license you acquire in accordance with your driving preferences.

Healing Tip 56

Humans need to love someone or something and need to be loved in return.
If you love your work, then you will love your life and your paycheck.
You can afford to love someone or something.
If you only love someone or something and do not love your job, then it will be half done, and your love will be undone, therefore you will not be loved.
Note: If you really do love your job, it will be a paid vacation

Healing Tip 57

Young kids talk about the future because they have no past.
Old people talk about the past because they have no future.
Healthy people (young and old) focus and work on the present in order to have a great future.

Healing Tip 58

Life Story of Mr. Right
When I was growing up, I noticed girls looking for Mr. Right.
I decided to find Miss Right.
After a long hard search I found Miss Right.
To my surprise after marriage she turned into Mrs. Wrong.
It took me many years of hard work to turn myself to Mr. Right.
If you are Mr. Right, anyone you marry will turn into Mrs. Right.

Healing Tip 59

Life story of Mrs. Right
I always wanted to marry Mr. Right.
I saw a handsome man and fell in love.
I asked his name, He replied Mr. Riot.
My love was not only blind, but also had defective hearing.
I heard Mr. Right.
My father told me "You sure have sweet taste for sour shit."
But I did not hear that; also I did not like it either.
Marriage was not only an eye opener; it also fixed my hearing.
I was healed. I gave Mr. Riot a permanent goodbye kiss and left.
I start living a healthy life till Mr. Right found me.
Now I am Mrs. Right and live a happy married life

Healing Tip 60

You are the product of your environment.

But you are the one who has to refine, reshape and make the finishing touches to yourself.

Be thankful to the environment for doing a great job. All you have to do is your job.

Healing Tip 61

You are the greatest gift to yourself.
Use the gift of your mind and do reasonable things.

Healing Tip 62

When the body is overfed or not properly fed, the human mind gets sluggish, and the soul starves.

The properly fed body means that the mind will be sharp, and the soul will gain external happiness.

Every meal you eat affects your eternal happiness.

Healing Tip 63

The thoughts I entertained and the one I should have entertained +

+ My actions and reactions; I chose and the one I should have chosen. =

The present mess I am in and more than my share of the universal mess I generated.

My thoughts of today + my actions and reactions of today, will determine my future in particular and universal future in general.

P.S. This universe is a unit, and I am a part of it. My thoughts, actions, and reactions not only determine my fate, but also affect the fate of the universe I occupy.

Healing Tip 64

To make the most of my life, first you must learn to focus on one thing at a time, and to make the most of present moment.

The present moment is the most important commodity you have.

Its use determines the quality and quantity of my life.

Healing Tip 65

On truly loving oneself...
My love for myself will be number ONE - I will know what love is.
My love for others will be SECOND to none.
I end up doing things...
I should not be doing and not doing things I should be doing, and justifying my mistakes instead of learning. I just keep making the same mistakes, and hoping for a better outcome.
How can I forgive and love all this?
It seems like mission impossible, but in reality it is mission possible.

Healing Tip 66

Everyone has his or her own unique shortcomings, transgressions and perfect hidden potentials.
To live in harmony with yourself and others, you must be able to see beyond the transgressions and shortcomings, and reveal the perfect hidden potentials in others.
It may seem that there are people more gifted than me and some less gifted than me. This is not true, in reality we are pieces of a puzzle, unique in our own way.
If I start serving everyone according to best of my ability, I may be able to see hidden perfect wisdom and learn from everyone.
I will be in able to see every piece in place and my picture will be beautiful and complete.

Healing Tip 67

Humans are in some ways are very similar to computers.
If you do not know my password or other's passwords,
You will never know the human's potential.

The only way to know your own password is to earn self-respect and to know other's passwords is to earn their trust.

Passwords change every day. To know new passwords every day, I have to stay up to date with continuing educational credits for self-awareness.

** If you know your password, you should never give my password to unauthorized persons**

If you forget your password or your password does not work, you can always use the universal password: GOD.

Healing Tip 68

Humans are meant to be complementary to each other.
$1 + 1 = 11$
Do not look up to anyone.
Do not look down to anyone.

Look straight into the eyes, and you will see your beautiful self in their eyes.

Healing Tip 69

Every human is a unique mirror with a unique frame.

When you see yourself in others, not only do you see your shortcomings, but you also see yourself in his or her unique frame.

It gives a new outlook and understanding.

Healing Tip 70

As I see it, there are four types of people- they all need each other to make a perfect world.

Type A) Healthy people they know themselves they are brutally honest, aware of their shortcomings and their past transgressions, and they work hard to improve themselves. They not only learn from their mistakes, but they are also able to relate to everyone else and learn from them.

They are courageous and motivated people.

Type B) These people know their shortcomings but lack motivation and courage, and are unable to live up to their full potential.

Type C) These people are unaware of their own shortcomings but very well aware of other's shortcomings and transgressions and use these to benefit themselves.

These people live comfortably unhappy lives

Type D) They have traits of Groups A, B, and C.

They should be the most responsible people and be role models for others.

Healing Tip 71

If you are looking for happiness, good health, and success,
You are wasting your precious time.
You must be honest with yourself and just do what you know needs to be done.
Happiness, good health and success will be looking for you!

Healing Tip 72

This day and age are the best because you live in them.
Capture the moment; the next moment is not promised and everything depends upon it.

Healing Tip 73

This universe is perfect.
There is mysterious, hidden, universal wisdom.
If you want to be unhappy, this is your choice.
In a less than perfect world you would not have that choice.
If you happy, you must remember you do have a choice.
In heaven the only choice you will have is to be happy.
In hell, the only choice you will have is to be unhappy
Enjoy my freedom while I am alive.

Healing Tip 74

Each human is equipped with reasoning program 101 since birth.

This program can be updated or deleted or distorted by the person in charge.

The generation gap is a reflection of different levels of reasoning powers.

Sometimes children outgrow their parents, and this is a called reversed generation gap.

Each human has a choice to update, delete or distort the program or stay unaware.

When was the last time you updated yours?

Healing Tip 75

Every human being has a purpose driven life.

There is not only a long list of purposes to choose from, but also a list of the qualities of purpose to choose from.

It is never too late to choose a purpose and the quality of that purpose.

Healing Tip 76

The people who abuse you are great blessings.

They can make you think, change, and grow, and also help you to appreciate the nice people in life.

If people or situations do not make sense, my senses are not mature yet.

A THANKFUL ENDING

Healing Tip 77

Life is a journey and there are three roads to choose from: Road #1 is uphill, Road #2 is straight and Road #3 is downhill.

The uphill road seems hard, but makes you healthier. Each step will elevate you to a higher level, and you will reach your destination.

The straight road seems easier, but has no possibility of growth.

The downhill road is the easiest and leads you into a hole.

You must know your capabilities before you try to help someone on a different road than your own.

Healing Tip 78

If you are trying to teach an unwilling old dog new tricks, you will become as frustrated as the dog.

However, if you are kind and understanding, then you can receive a lot of wisdom from that old dog.

Healing Tip 79

If you pay attention to everything that happened to you, then you will see perfect hidden wisdom in your life.

It does not matter, good, bad, pleasant, unpleasant, humiliating, rewarding, painful or comforting, if anyone respected you or was unjust to you. Everything and everyone should remind you that you can learn.

Healing Tip 80

If I do live a purposeless life, then my life will be money-less, happy-less and everything-less.

If I live a purposeful life, then I will have a happy-full, success-full and full life.

P.S. If we choose to live a purposeful life then we will have lived a full life.

Healing Tip 81

All humans are created equal but different.

Each human has his or her own strength and shortcomings, like different pieces of a puzzle.

If all pieces are placed in their proper place, it will mirror the perfect universe.

If you find your proper place, you will serve your purpose and live a perfect life.

Healing Tip 82

My life is a paid vacation.

My mind and body do the work, and I get the paycheck.

P.S. I must take of my body and mind; otherwise no paycheck

ADDENDUM: My body is like a goose; when it is healthy, it gives golden eggs; if unhealthy it may end up having diarrhea.

Healing Tip 83

The quality of the present moment will determine my future.

If the baggage from my past or worry about my future is affecting the quality of my present moment, then I am not living up to my full potential.

P.S. The correct diagnosis of my problem is most important. If the problem lies in the past or future, it must be addressed.

Otherwise, I will never have a perfect moment and perfect future.

Healing Tip 84

I not only had learned a lot from others, but I also kept learning every day.

What I have learned by making dreadful mistakes, barely surviving and recovering this firsthand experience is no match to any other way of learning.

Whenever I see someone or myself in a dreadful situation,

I know it can be a great learning experience if I am able to overcome it or try to help someone else to overcome it-I will become a better person.

ADDENDUM: If you acquire a dangerous disease survive, not only will you become immune, but you will also be motivated to invent a vaccine for others so that they will not get sick. (Hidden perfect wisdom)

Healing Tip 85

33 Miners are rescued; everybody seems happy.

What the future holds for them is how their brains process this event.

Three ways:

1) Process as a reminder and become better and healthier people.

2) Try to forget, and keep going on with life.

3) Go through post-traumatic stress syndrome, depression, nightmares and become unhealthy people.

I see it as a reminder to humanity to work together, to be happy for each other and to not wait for another reminder that may not have a happy ending.

I am feeling the taste of heaven in this life, and I wish every human will feel this heavenly feeling.

Healing Tip 86

My father was a very wise man, and I learned a lot from him.

When I was ten years old, I noticed that he worked hard, came home late and ended up having leftovers. I felt that was not fair.

I told him my concern; I felt he deserved a decent meal.

He appreciated my concern and replied "What can be done?"

I felt mother should keep his meal before serving others.

He asked if he said something to give me this idea.

I proudly replied that I had thought of all this by myself.

He said what is wrong with this picture is that my mother is raising an ungrateful son like me, who has no respect for his mother today, and tomorrow I will have no respect for my wife.

He assured me that my life will be doomed.

He advised me to respect and serve my mother, and what is between him and my mother is none of my business.

I was shocked and I thought this is a crazy man, who deserved leftovers.

Healing Tip 87

I saw God in my dream.

God asked me "Tell me something you did that you are really proud of."

I asked God "You tell me what you did that you are proud of."

God replied "I made humans. They are my masterpiece."

I was surprised; I said pointing toward me "You call me your masterpiece?"

God replied "Yes, this unkempt masterpiece has an unhealthy junk piece. Do what you know you must do; then you will understand what I mean."

I had a lot of questions; I woke up confused.

After many years I realized that I must do the healthy things, and stop doing unhealthy things, and clean up my mess.

I will turn back to be God's Masterpiece.

Healing Tip 88

Humans are like trees.

They have a choice to be fruitless or fruitful.

If they choose to be fruitless, they may serve some purpose, and may be used as fuel paper.

Or, they have a choice to be fruitful and give a constant supply of fruit and serve a better purpose.

Fruit bearing trees like fruit bearing humans never die because their seeds continue to grow and make more trees and live forever.

Healing Tip 89

Your eyes, lips and ears must be sealed.

Stay connected to the source, and be able to see and hear perfection in everything

Healing Tip 90

All my life I was looking for someone to help me find my destination.

I followed many people and ended up being disappointed every time.

Finally, I discovered I am the traveler, and I am the destination.

Keep both eyes on your destination and start this long most challenging journey. Just knowing that when you reach your destination, it will be the most rewarding moment of your life.

Keep both eyes on the destination, keep going, and no distractions.

Healing Tip 91

Moving water always stays fresh. (Healthy water)

Stagnant water is shaped by the container and becomes toxic (Unhealthy water).

Thinking minds shape the world. (Healthy mind)

Closed stagnant minds intoxicates the world. (Unhealthy mind)

Healing Tip 92

Your body is like a car.
Your brain is like a driver.
Conditions of the car reflect the driver's state of mind.
Condition of your body reflects your body's state of mind.

You have a choice to do regular recommended maintenance on your car.

If you choose not to, you may have to call a tow truck or call 911.

If you still do not learn, you have to call the junk yard or someone will call the funeral home.

You must do tune ups, oil changes, and heavy duty cleansing of your mind in a timely fashion, before it is too late.

Healing Tip 93

Once upon a time a man was sitting on a tree branch while cutting the same branch.

A wise man was passing by and advised him not to do that.

The man advised him to mind his own business.

The wise man apologized and gave him $10 for his advice and left.

When the branch was cut the man fell on the ground.

He ran toward the wise man and asked him: "You must be a magician, how did you know ahead of time that I will fall? I would like to learn that magic from you."

The wise man replied "Sure that is not a problem, you taught me so much and I owe you a lot." They became good friends and kept learning from each other and were a compliment to each other for the rest of their lives.

Healing Tip 94

I have learned so much from my children.

My daughter when growing up used to demand me to tell a story; in which a female must be the king.

I made up a story of Deer King.

A long time ago there was a middle age female deer struggling to survive with her five kids.

Her husband was eaten up by the Lion King.

One day she decided to stand up against the Lion King who eats up innocent citizens.

She started working and found a lot of other animals felt the same way, but no one had the courage to do anything about it.

Very soon a lot of animals were her supporters.

They declared her king of the jungle.

Lion King was furious and came to kill her.

She was prepared, he tried to catch her, but she was fast.

The Lion King was huffing, the deer jumped over a ditch she had especially prepared. She jumped across the ditch, Lion King followed her, but was not able to jump across the ditch and fell into the ditch. Now he was the prisoner of the deer.

She was the Deer King of the jungle.

She made the constitution based on fairness, equality and justice.

There was peace and harmony in the jungle and everyone lived happily ever after.

Healing Tip 95

All my life when I looked up to someone, I was not able to see straight, and ended up falling on my face. (Moral: Never look up to any human being)

All my life when I looked down to someone, I went down. (Moral: Never look down at any human being)

All my life when someone pointed a finger at me, instead of being thankful to them for showing me my shortcomings- I disliked that person. (Moral: I must appreciate people for pointing out my shortcomings to me, which I cannot see myself. I must be thankful for their free services.)

All my life I always had a problem; I finally found the solution within me. If I have a problem, I have the solution. (Moral: I must first look within me for the solution for my problem)

Healing Tip 96

If I am a healthy human being, I will be able to coexist with everyone, not only without problems but will also benefit from everyone.

P.S. All I have to do is to stay in good health and keep a safe distance. It will be heaven on earth.

Healing Tip 97

When a courageous person speaks his or her mind, it sometimes may cause trouble.

A healthy wise person not only can read his or her mind, but also can read other people's mind, always a peace maker.

Peace is healthy, trouble is unhealthy.

Healing Tip 98

Life is like a game of chess.

If the goal is to learn and to get better, learn to focus on yourself, and learn to keep yourself safe by putting your pieces in the right places.

Learning and having fun is more important than winning. Become the master of self-defense.

You may lose some games, but you will be victorious in the war of life.

A Lovely Ending!

Healing Tip 99

Sick body hurts.

Sick mind's pain is worse.

Pain of a sick soul is unbearable, if it is still responsive.

Pain is a great friend and reminds you to take care of the root cause and become healthy again.

If you keep ignoring the pain or have no pain, you will lose my soul first, then the body, then the mind.

Healing Tip 100

You are the book.
You are the writer.
You are the main character.
This book has three chapters.
Chapter #1 - Past
Chapter #2 - Present
Chapter #3 - Future
Choose the characters carefully, and get rid of old characters who no longer have anything to offer.

Thank them and give them a warm goodbye.

Always have the end in mind, so you do not get distracted.

This is the end of the Trail of Healing 101 - Lesson #1

If you have successfully climbed the trail, proceed to Lesson #2

Healing Tip 101

The human mind is the most sophisticated computer known to humankind.

Healthy human mind can process all self-shortcomings and ability to delete and trash them.

Healthy human mind can inter-phase with everything and everyone, download healthy traits and useful information, recognizing and blocking the negative traits and toxic information and stay on top.

A healthy human with a healthy mind can live a peaceful, happy purposeful and successful life, learning from everyone and sending positive messages.

A human mind if not maintained and updated according to the manufacturer's recommendations, will malfunction and create disaster for self and others.

Healing Tip 102

A healthy heart does not harbor any hard feelings toward anyone, anything or toward himself or herself.

A healthy heart is full of self-love and love for everyone and everything.

Healing Tip 103

Each human is unique (like fingerprints) and everyone will be provided different directions depending upon his or her condition.

WARNING: Do not follow others' directions or criticize others to try to tell others to follow your directions. It will create disaster and all guarantees will be void and null.

Healing Tip 104

Healthy reasoning is the most powerful force second to none.

A human cannot force himself or herself to change.

Any such attempt will lead to civil war within.

Use of force without healthy reasons is very destructive.

It always ends up in a lose-lose situation.

You can change any habit or trait, regardless how deep-rooted it may be, by enough powerful healthy reasons.

It will always be a win-win situation.

Healing Tip 105

Logic of nature is very simple.

There are problems, and there are hidden solutions, waiting to be found.

There are diseases, and there are cures, waiting to be discovered.

Humans have multiple choices.

1) Live life and not worry about problems, and diseases just go with the flow.

2) Live with problems and sickness: just keep complaining.

3) If acquire problem, seek for solution, if get sick, look for cure.

4) Be self-aware, constant self-evaluation, and nip the problem and sickness in the bud, and live a happy and healthy life.

5) Be not only self-aware, but go above and beyond, acquire wisdom, notice that humanity demands solutions and cures, study problems and illnesses, do research, learn from everyone, discover nature's mysteries, discover solutions and cures, so everyone benefits from them.

Healing Tip 106

I can learn from myself.

And I can learn from others.

Others can learn from me.

No human has the ability to teach another human being.

Every encounter I had in my life, I learned from everyone, so everyone is my teacher.

I must be thankful to them for their great favor.

Healing Tip 107

Each human is a character on the stage of life.

Each human also has characteristics (some inherited and some acquired).

Each human has a choice to either find or write a script and follow it, or live without a script, and nature will take its course which may be undesirable.

They need this script to practice and play their role well.

Determination needs to be acquired to live up to their script.

Humans are actors and writers.

Healing Tip 108

I am fascinated by nature's wisdom.

I look back and realized that everything had wisdom within that was hidden to me at that time.

Today I can see perfection in my past and present, and I am sure whatever will happen in the future will be perfect.

You always have a choice to see or not to see.

Acquire clear vision.

Healing Tip 109

Yesterday I attended Thanksgiving services at Dauphin County Prison, Harrisburg Pa.

All inmates were invited regardless of their background.

It was a true human service lead by Chaplin along with Rabi, De con and Imam.

This was one of the most powerful Thanksgiving services I ever witnessed.

All humans regardless of their background under one roof giving thanks for their blessing.

It was my dream come true.

Humans are just one family.

I felt the grace, purity and saw that inmate's blessings (because they are serving) and my opportunity to earn blessing by serving them.

Happy Thanksgiving to all Humans!

Why can't that kind of Thanksgiving Service be done outside of the prison?

Healing Tip 110

Health is the second most important gift, after faith, to be thankful for.

Healthy = Happy

This is the story of three unhealthy men, 70 years, 80 years and 90 years old.

The 70 year old man was complaining that being 70 is hell, "I cannot hold my water and it is embarrassing."

The 80 year old man said "You have no reason to complain, I should be complaining because I not only cannot hold #1, I cannot even hold #2 - it is a mess."

The 90 year old man said, "Why the hell are you two complaining? I should be complaining, I lose #1 at 7am every morning and I lose #2 at 9am every morning - and I wake up at 12 noon."

That is a real mess.

Moral of the story: Take care of your health, otherwise you will always find something to complain about.

Healing Tip 111

Never try to teach someone without learning from yourself first.

Of course everyone is welcome to learn from me for free.

Healing Tip 112

Balanced life means 100% focus on present.
Perfect foundation (PAST) must be in place.
Today will be tomorrow's yesterday another brick to the solid foundation for a great future.
You must start digging to find and fix the defects...
CAN YOU DIGG IT?

Healing Tip 113

I had trouble coexisting with myself, so I became codependent.
Everything and everyone I depended on proved not to be dependable.
I had to come to terms with myself.
Now I have peace with myself.
I cannot only coexist with myself; I can coexist with everything and everyone.
Find peace within yourself, before it is too late.

Healing Tip 114

I had learned with my life experiences, that it was abundance of resources that made me mostly lazy and dysfunctional.

It was too many choices; mostly made me pick the wrong choice.

Lack of resource motivated me, made me feel free. I became creative and I became resourceful.

Lack of choices, forced me to create my choices and successes.

It is not lack of resources or lack of choice, it is the lack of motivation that is the root of the problem. Lack of motivation is usually caused by abundances of choices and resources.

Healing Tip 115

Humans are captive of their attachments.
Make sure you have pure attachments for the right reasons.

Healing Tip 116

When you finally get in touch with your feelings, you will discover you harbor many feelings that you were not aware of. They might have an impact on the quality of your life.

They could be lumped into three categories, Good, Bad and Toxic.

Hard feelings toward yourself and others were like toxic waste.

The world would be a better place without toxic waste.

Healing Tip 117

Good health is supported by five pillars.
Quality of pillars determines quality of life.
Pillar #1 - Healthy Diet
Pillar #2 - Regular Exercise
Pillar #3 - Good Night Sleep
Pillar #4 - Earning an honest living is the best form of worship.
Pillar #5 - Must have down time of your choice
Pick up one or more than one from the following choices:
A) Prayer
B) Meditation
C) Quiet time
D) Custom made down time.

Healing Tip 118

The key to success is to be the right person.

Every human touch is the most powerful experience of life; it can be positive or negative depending upon the state of human mind.

With proper state of mind every encounter adds a new dimension of human understanding.

Healing Tip 119

A car, computer and engine will not perform 100%; unless maintained properly according to recommended instructions.

Human mind and body is no exception!

A car, computer and engine come with manual and instructions; it must be followed for best performance.

If I do not have a manual to follow, nature will take its course, your primitive mind will take over, and you will not live up to your full potential.

If you take charge and make your own manual, you will live a full life and will have control of your mind.

Healing Tip 120

A healthy heart is a pure heart.

Free from toxic substance and toxic waste like hard feeling, ill feeling, toxic love for self and others, and any kind of negativity.

A pure heart is also sensitive enough to recognize and block any unhealthy material, as well as able to recycle to healthy and positive material.

Healing Tip 121

Each human has four common traits.
1) Ignorant dominant (do not know but think they know it all, do not feel the need to seek knowledge.)
2) Unaware dominant (has incorrect understanding but are sure that what they know is correct and are eager to teach everyone, not receptive to learning.)
3) Awareness dominant (seek knowledge, learn from his or her own past, learn from type #1, type #2 & type #4 people and live a happy and successful life.)
4) Mixed kind (keep changing with time and circumstances, their life is like a roller coaster (up and down depending on the moment)

Each human has the ability to transform into an aware dominant person. It is never too late.

Healing Tip 122

The time train moves on, the present moment will become the past moment, the future moment will become the present moment and so on.

Every moment is a choice for you to be a peacemaker, troublemaker, spectator or anything you like to do with that moment.

How you use every moment will affect nature's course in one form or another.

You must pay attention to how you utilize every moment.

The future of humanity depends on it.

Healing Tip 123

This world is a great human lab.
We can learn about how nature takes its course.
If we study present history, past history, others' history or our history, and learn how people changed the course of history for better or worse – we can conclude what role we must play, and change the course of nature, and be a history maker.
Human resources are unlimited.
Each human alive or dead is a great treasure, just waiting to be discovered.
Every human is making a history, whether they are aware or not, for better or worse. We all have a choice.

Healing Tip 124

Religion is just a tool.
Some healthy people use religion to become righteous.
Some unhealthy people use religion and become psychotic.
It is not the religion; it is the person who practices it.

Healing Tip 125

Forgive and forget sounded good and I practiced it, until I lost the capacity to forgive, and forgot about forgetting.
You must first forgive yourself.
Do not forgive yourself twice for making the same mistake. Instead, repent and learn the lesson again.
First learn, then forgive and never ever forget the lesson.

Healing Tip 126

Someone told me "when the student is ready – the teacher arrives."

All my life I not only waited for the teacher, but I kept looking for the teacher.

No teacher led me to the destination I wanted.

My life continued to get worse instead of getting better.

I stopped looking, closed my eyes and saw my teacher within me.

I not only did not like the teacher, but I did not like the teachings either.

But I did not have a choice.

If you start listening to your teacher, you will learn to love them, for they are with you 24/7.

You are the teacher, student, and destination.

All other people are substitute teachers.

MORAL: Stop looking, close your eyes and meet your real teacher, then you can start learning.

Healing Tip 127

A knife in the hand of a skillful surgeon can save a life, but in the hand of an inexperienced surgeon, it can take a life. In the hand of a child it can be dangerous, in the hand of a criminal it can kill few people, and in the hand of a psychotic religious person (for example a terrorist) it can be a weapon of mass destruction.

Deal with your weapons carefully.

Healing Tip 128

Many people deal with challenges from their enemies successfully, but not peacefully.

You are your own worst enemy. If you treat the struggles with yourself peacefully, you cane become your best friend.

Healing Tip 129

There are three kinds of people —
1) People with predominately happy memories. They seem happy most of the time.
2) People with predominantly painful memories. They seem unhappy most of the time.
3) People with no memories, self-centered people. They seem the same most of the time.

Painful memories force you to look deeper into yourself and can eventually help you grow.

They are like manure, in a fertile land, with good seeds it will turn into beautiful fruits and vegetables; otherwise it will be a very foul smelly life.

Healing Tip 130

Old age is a golden age, it is true.

A healthy old person, keeps control and lives a happy golden life.

An unhealthy old person, loses control, and will need assistance.

Stay healthy, stay in charge, and keep your control or else.

Healing Tip 131

A healthy person is one strong unit like NO 1.

A healthy person is flexible and is in position with everyone appropriately. With another healthy person it is 1+1=11.

An unhealthy person is negative.

A healthy person not only coexists with everyone, but he or she also benefits from everyone.

Healing Tip 132

Happy moments will turn into happy hours.
Happy hours will turn into happy days.
Happy days will turn into happy golden years.

Healing Tip 133

Life is a book.

Every day's a new chapter.

Every morning is the beginning of a new chapter.

Every night is the end of the chapter that very well could be the last chapter.

Every day must be a better day than before, and the final chapter must be the best.

End of the chapter.

Healing Tip 134

Blame yourself for the mistakes you have caused.
Blame others for the wrongs they have done to you.
But if you give credit to yourself for all the right things you have done, do not forget to give credit to the others for being the reason.
It does not matter if those things were good or bad.
Nature has hidden wisdom in everything which happens, waiting to be discovered and understood by humans.

Healing Tip 135

There is a saying that goes something like this "Aim for the moon, if you miss, you will hit the stars."
"Aim high, and you may land much higher than ever expected."
Do not attach yourself with the outcome, stay free, set no limits, and always do the best you can. This will lead to unlimited success.
Human potential and abilities are limitless. Use it wisely and pleasantly surprise yourself and others.

Healing Tip 136

Yesterday I had a very significant encounter as usual.
I met an old friend with a female companion.
I asked her "You must be his significant other."
She replied "I am his insignificant other."
I asked "So you are not significant?"
She replied "I do not think so."
I could relate to her because I used to feel the same way.

Over a period of time I learned that I am the most significant person in my life.

Since then every encounter became very significant, reminded me of my significance and helped me grow.

Everyone treated me the way I treated myself.

I learned to respect myself in order to be respected.

I live by "Love others as I love myself." I learned to love myself and acquire the capacity to love others.

Others who try very hard to prove that you are just an insignificant other can teach you very significant lessons.

Healing Tip 137

A long time ago there was a princess, who was also a great singer.

One day she saw a cute pig and they instantly fell in love with each other.

The princess was dreaming that she would teach the pig how to sing, and that they would be stars.

The pig was dreaming that she would be a great mud buddy and live like a pig.

The princess tried to teach him how to sing, but both became frustrated. The princess gave up and let him live like a pig.

The pig starts teaching the princess the benefits of being a pig.

Her love was blind, so she becomes a pig, and they both lived together in a pig house.

If you fall in love with a pig, make sure that your love is not blind.

LOVE WITH OPEN EYES IS TRUE LOVE. IT CAN MOVE MOUNTAINS!

Healing Tip 138

I want to take a moment to thank everyone in my life especially my patients who are not only a source of inspiration, but also a source of revenue, including everyone who made valuable comments to help me be the way I am.

Make sure to thank those who you are close to.

Healing Tip 139

When I was in Medical School, one of the professors was emphasizing the importance of listening and being able to relate to a patient.

He said "A good doctor can tell that the patient is sitting in front, the quality and quantity of a patient's life and possible cause of a patient's death. Also recommendations that will improve the quality and increase the quality of a patient's life."

I thought at that time that it was farfetched. But today I know what he was talking about.

If the doctor is ready to listen, and gain the patient's trust, then the patient will tell the needed information to determine the patient's condition.

The doctor must be ready to listen to the patient's concerns and remind himself the multiple benefits doctors will earn.

It will be a hard and long road, but physicians can be a great catalyst to the patient and both patient and physician will be surprised with the outcomes.

First class medicine is a prevention of disease, Second class medicine is treating the illness, and third class medicine is treating the complications.

Real medicine is to listen, relate and and accomplish mutual learning.

Healing Tip 140

I have heard many versions of this story, and this is my version.

A physician with a rescue team was visiting a flood affected area.

He saw a scorpion on a floating branch ready to drown.

The physician tried to rescue him, but every time he tried to attempt, the scorpion attempted to sting him. This process continued for a while.

One team member said to the physician "Leave him alone, what are you doing?"

The physician replied "The flood is doing what the flood must do, and the scorpion is doing what he must do, so I am doing what I must do."

Lessons of the Story:

Natural disasters always have hidden wisdom to be understood and in some cases are reminders to humans to change their unhealthy behavior.

Life in any form or shape keeps a balance in nature and must be preserved.

There are always opportunities to learn and do well in any circumstance or encounter and to be part of the solution.

Leaving things alone and walking away could be a missed opportunity to learn and grow.

PAY ATTENTION TO THE BIGGER PICTURE!

Healing Tip 141

I heard this story a long time ago, but today, it seems like my life story.

The principal was inspecting a school, and asked a student what is 4+4.

The student answered 16.

The principal looked at the teacher, and the teacher said "He came a long way he used to reply 44 - I hope one day he will be 100% correct."

It does not matter how long the journey takes you, only that you continue to make progress on the way.

Healing Tip 142

LAW OF NATURE: Every action has its reaction, and every reaction has its consequences.

Healing Tip 143

Self-abuse (physical and mental) in any form or shape is a serious crime in nature's court, punishable to unhappy, stressed life imprisonment and/or prolong miserable death sentence depending upon the nature of the crime committed.

Four out of four humans suffer from self-abuse.

Do you know how to be safe?

Do you know where to get help?

Abuse is not really your fault.

Call toll free 24/7 **ALL CALLS ARE FREE AND CONFIDENTIAL** 1-800-CRE-ATOR - Use heart phone or mind phone only for identification and download your custom made program.

YOU WILL RECEIVE YOUR MANUAL OR HAVE IT INSTALLED IN YOUR MIND OR HEART INSTANTLY.

UPDATES ARE AVAILABLE FREE OF COST!

Healing Tip 144

There is a saying "Easier said than done."

It is only true is you say it, but do not mean it.

Having a plan and declaring the intention are the hardest parts and foundations of any task.

Healing Tip 145

In this well balanced world, there are the right numbers of nuts and bolts.

Quality of this world depends upon the number of loose nuts and wasted bolts, more loose nuts and wasted bolts, equals more chaos in this world.

Each nut and bolt is crucial, if each one is placed properly, this will be heaven on earth like pieces of a puzzle placed properly it will complete the beautiful picture.

Healing Tip 146

My father was a very unusual person.

He had his principals. He also never fit properly into society, and was considered not very successful according to the standard of that time

When I turned 20, I confronted him, and questioned his behavior.

As usual, instead of directly answering, he told me this story:

Once upon a time a monkey accidentally got his tail cut off.

It was painful and difficult jumping on trees with no balance.

He was afraid he would be an outcast; a monkey without a tail would not fit into a respectable monkey society. He decided to talk another monkey into cutting his tail off, convincing him that having a tail is old fashioned. "Look at humans," he said, "they are more advanced than us because they do not have a tail!"

It was a painful experience and now the other monkey could not jump or keep his balance either.

He started cursing the other monkey for talking him into cutting off his tail.

The first monkey said "If you tell other monkeys, we will be outcasts and they will think that you are a fool. Let's start talking everyone into cutting off his or her tails."

Very soon monkeys with tails became the outcasts.

However, one day, one monkey declared that having a tail is normal, he went through hell, but his walk was louder than his talk, and one day all monkeys decided they wanted tails again.

Since then, all monkeys are living here after with their tails.

I could not get the message then, but now I understood what he meant.

It can only take one person or being to change how everyone else sees the world.

One happy person can make the whole world happy.

Healing Tip 147

There are three types of people.
People who have the problems,
People who are busy creating the problems,
And people who are busy solving the problems.

If the 3rd type of people stop bitching about type #1 and type #2 people, appreciate them, love them, and get to work, they will always be ahead of the game.

Very soon people who create the problems will not be able to keep up and give up.

In that case, type #3 people still must love them in spite of the fact that they are not giving them enough business.

Then type #2 and type #3 must concentrate on type #1 people to help them, and still there will be enough business for everyone.

Healing Tip 148

It is not what is being read, but instead it is how it is being understood.

There is great wisdom in this statement, but of course, it can be understood differently.

Healing Tip 149

Religion without righteousness can be very self-destructive, and it also can be a weapon of mass destruction.

Righteousness with or without religion can help to gain self-peace- peace with everything and everyone.

Healing Tip 150

Every human is a treasure.

Every encounter is an opportunity to learn and grow.

Someone once said "If a doctor listens to patient, patient will tell what is wrong with him or her. If a doctor listens more carefully, patient will tell what is wrong with the doctor."

LET YOUR PATIENTS BE YOUR HEROES!

Healing Tip 151

Your mind and body both need healthy food.

If I feed my mind first with healthy food, then I will feed my body with healthy food.

If I ignore my mind and feed my body, then I will feed my body unhealthy food, and end up having an unhealthy body and starved and sick mind.

Healing Tip 152

Can you visualize your future?

What will you be when you are 90?

You must do what it takes to stay healthy.

Then you will be 90 living independently, enjoying life, and helping others, not an unhealthy wheelchair bound 90. It all depends on if you make the right choices while you still have time.

Healing Tip 153

Every human being has a choice.

To have a peace treaty with self, and live a happy, healthy and purposeful life.

Stay in a state of undeclared war by doing unhealthy and harmful actions against yourself. If this undeclared war continues for a long time, then some of the damages become non reversible, heading for slow lingering painful death. In case it becomes overwhelming, this may lead to successful suicide.

It is never too late to create a peace treaty with yourself.

Healing Tip 154

There is an old saying.
Best doctors prevent disease.
Doctors treat disease.
Specialists treat the complications.
Life is like a chess game. To win I must have a great defense.
I will wait for the other person to make a mistake.
The one who waits and has a better defense wins.

As far as my health is concerned, it has two defense mechanisms.

The first line of defense is not to allow any stress, anger, jealousy or any negative thoughts and feelings, any harmful substances, bacteria, virus and guard all ports of entry (what goes in and what goes out) including what enter and exit out of my brain with my life.

The second line of defense kicks in when the first line of defense breaks; try to fight back the invader (toxic food, thoughts or any harmful thing) and most likely overcome the problem; but if this process continues, first the quality of life starts affecting and later starts losing the quantity of life.

As long as you are alive, it is never too late. You can recognize your priorities straight and work on your defenses to regain control of your life.

Healing Tip 155

There is an old story that a man went to Moses and asked him "You talk to God all the time, if you please, ask if I am a good person."

Moses replied "If you don't know yourself, then ask your wife."

Healing Tip 156

The glass is half full or half empty.

If I see it half full, then I will be thankful of what I have, and I will enjoy as it energizes me and very soon my glass will be full.

If I see it half empty, then I will be unhappy, jealous of people with full glasses, and not able to enjoy what I have, therefore becoming miserable and ending up having an empty glass.

The glass is really what you make of it.

Healing Tip 157

Self-awareness is the key to live a life.

The human mind is driven by limbic system and/or cortex.

The limbic system includes the primitive brain and responds to impulses, temptations and learned behavior.

The cortex is the higher center that gives humans reasoning and consequences.

The limbic system looks for instant gratifications.

The cortex reminds us of consequences and offers healthy alternatives and options.

Limbic driven people are usually in trouble.

Cortex driven people live a happy and balanced life.

Healing Tip 158

I may be alone but I am never lonely.

I always was, am and will be in a world, it may be a fantasy world, real world, dream world, evil world, hostile world, past world, present world, difficult world, in my own world, friendly world, or any other world.

Regardless where I was, where I am or where I will be, if I learn from every world, then I will discover my perfect universe.

Create a world where you never feel lonely.

Healing Tip 159

I dedicate this Healing Tip to my wonderful friend Carl from Camp Hill Pa, who inspires me to write this healing tip - Thanks Carl!

Having realistic expectations is a key to a happy and successful life.

If you invest in yourself, good things will come.

I will receive more than I ever expected.

If you invest in others, it is a risky business.

Even if you are good to others, you must be aware of their worst potentials.

Always be prepared for the worst.

Others may not do anything in return for you, but you need to accept that.

When they do, you will be pleasantly surprised.

Invest in yourself and you can afford to take a calculated risk to invest in others and grow.

Healing Tip 160

I dedicate this Healing Tip to a wonderful caring nurse Jane from Camp Hill Pa, who inspires me to write this healing tip. Thanks Jane!

You can learn a lot from your friends and teachers, but even more so from your so called enemies. They force you to learn.

I being my worst enemy, I must learn most from myself.

Healing Tip 161

I dedicate this healing tip to my patient who does not want to be named. I will call him Joe Smith - he inspired me to write this healing tip. Thanks Joe you are my HERO!

I saw him many years ago, and he was having a lot of problems.

He said "Everyone thinks I am a zero and everyone treats me badly. My life is horrible."

I asked him "What do you think are you a hero or a zero?"

He said "Definitely a zero."

If you know you are a zero, then you must position yourself correctly. For example, 1+0 = 10. That is a proper position. However, 0+1 = 01. This is an improper position, which will only cause problems for yourself and others.

I told Joe to put his wife first, because HER + 0 = HERO. He can make her feel like a hero, but if he puts himself first, it only becomes 0HER, which is nothing but trouble.

In order to become a hero, first consider putting yourself in a better position so that when you are added with others, you no longer appear to be a zero.

Healing Tip 162

I dedicate this healing tip to my father who was a great spiritual being. THANKS DAD!

An engine has 3 basic needs: 1) Gas, 2) Tune-Up & 3) Regular maintenance for trouble free performance.

A human being has 3 basic needs: 1) Healthy food, 2) Regular exercise & 3) Good sleep to live up to full potential.

Do the basic that is the foundation to live a full potential.

Healing Tip 163

People die the way they live their life.

Being a doctor, I had witnessed a lot of deaths.

None beat the experience I had when I was 20 years old and witnessed the death of my friend in Quetta, Pakistan.

I was born and raised in a small town of Pakistan - Quetta.

There was a shop owner who used to repair and sell watches.

He was rather strange, had no family, no friends - a loner.

There were lots of rumors about him. Some would say he was a spy, others thought he was an undercover agent of some sort.

Once, one of my neighbors got in trouble at work because of his defective watch.

He was too poor to afford to repair it or buy a new watch.

I took his watch to that shopkeeper, who repaired it and asked me if it was my watch.

I told him that it belonged to a poor man who could not afford to get it fixed, so I was doing it for him.

He said he would cover the cost, and we became friends.

I discovered he migrated from India after partition in 1947.

He was eight years old at that time.

His whole family was killed, he saw the dead bodies of all his family, and he was the only one that escaped.

He was raised in orphan homes and ended up settling in Quetta, and decided that he would stay single and treat everyone like his own family. He was a major contributor to orphanage homes.

He lived in a small very simple rented room.

I was very much impressed by him.

He then got sick and I was the only one to take care of him.

He was in a coma and I heard him saying clearly, "O'GOD I beg for forgiveness. If I have done anything wrong to anyone bless them so much that they will forget the pain I caused. If someone has done wrong to me, I am forgiving them and beg you to forgive and reward them. I beg for forgiveness of all my sin and I am thankful for the ability you blessed me to do any good." He died peacefully.

I was stunned. I really felt a nice breeze and saw white light and peace I never felt before. An experience of a lifetime.

If you want to die peacefully, you must try your hardest to live a peaceful death.

Healing Tip 164

If someone hits you once, you might ignore it.

But if it becomes a routine practice, then you should fight back and attempt to stop them.

If you are your own enemy, then stop yourself.

Healing Tip 165

This world is made to be explored by humans.

A human's mind and body is the most sophisticated system complex with unlimited potential.

As a human, I am president: in charge of my system complex.

If I choose to entertain negative thoughts, harbor hard feelings towards myself and others, or act and react in an unhealthy way, then my system will sense that and provide me unlimited constant stream of the same negativity.

Not only I will hurt myself, but I would have a negative universal impact.

If I choose to entertain positive thoughts, have good feeling towards myself and others, and appreciate everything and everyone and act and react in healthy ways, my system complex will provide a constant stream of the same positivity.

I will continue to grow and become a better person, and I will have a positive universal impact.

Healing Tip 166

Tell myself every day that I do not have to be depressed; I can find a way not to be depressed. I will come out of this knowing how strong I will be, and will not think of it as the hardest thing I have to do.

Healing Tip 167

MY PATIENTS ARE MY HEROES.

I do the talk and they do the walk.

I do not care what people think about me, as long as I live the right way, treat others as I want to be treated. First I must learn how I must treat myself.

I have a goal I must achieve. My training period is hard and lot of work. I am not going to let things upset me. I know if I complete my training period successfully, I will have a great job. I must focus on my goals.

I will not let anyone put me down anymore, what they say is none of my business.

I will get through this and will be a stronger person. I will be strong and independent. Good things are happening to me, and I must be strong to take advantage of these opportunities.

I will continue to evolve to be a better person every day. I know I need help. I must take my medications as prescribed. I must not stop taking medication without discussing with my doctor. I know it took years to get here, I must not rush, and things will not get better overnight.

Healing Tip 168

I always wanted to be a doctor, but a lot of people told me I was too stupid to become one.

My father was the only one that told me otherwise.

This was true because not only I was not able to concentrate on the tasks, but it used to take me much longer to complete my tasks than others.

I had accepted the fact that I was stupid and was content with that.

But, my father never gave up on me, and kept working on me.

Of course I could not get into med school due to my disability, but I got into Pharmacy school.

It was a real struggle; it used to take more than double the effort even to pass.

I used to question why everyone was having a good time and receiving better grades than I.

With my father's support, I made it through Pharmacy School and ended up in the USA.

I found wonderful people, who helped me not only to get a Pharmacy License in the US, but also encouraged me to go to med school.

To my surprise, I not only became a doctor, but I became a really successful one.

To me the US is a wonderful place where anything can happen.

Have a dream. Search your soul, you will discover the real you and surprise yourself and others.

Healing Tip 169

Human's actions and reactions are either feeling driven (heart) or driven by reasoning (brain).

Some say that women are predominantly feeling driven, and men are driven by reasoning, and that is the reason they do not get along.

In my case I use my feelings as a primary source, and when feeling does not work well, reasoning kicks in and vice versa.

Healthy reasoning and healthy feelings are a complement to each other to live a happy, healthy and successful life.

Feeling and reasoning must work as one unit, and to do so it needs constant work and maintenance.

Healing Tip 170

Humans are the supreme creation and have three choices:

To have the potential to be Supreme Being and live a pure purposeful life and blessing to everyone with awareness and righteousness.

To to stay healthy and live like a human being.

To do evil, be self-destructive and hurt others and live like a devil being.

These paths are decided by your actions and choices.

Healing Tip 171

Every sinner has a future.
Every saint has a past.
If you capture the present moment and make the most of it, your past will turn into a treasure, a great present, and bright future.

Healing Tip 172

On May 21, 2011, I was awarded the John Schultz Compassion in action award by Blue Lotus Society, celebrating the birth of Buddha in Harrisburg!

I was asked to speak and this is the text of my message:

Every human needs faith to function, so do I.

I must be self-aware and do myself purification for the best result.

ME + My Faith = What I am.

If I am a defective product this means I have not done my homework properly.

I know the universe is perfect and there is perfect wisdom in everything and everyone.

I also know that everyone and everything is serving his or her purpose.

I must know my purpose and do what I must do.

All my life, I felt that something was missing.

Now I know that with self-awareness and self-purification, I can discover my heaven within me. I know I have a long way to go.

When I meet my Creator, I will be asked if I had everything to discover my heaven within me, and I was not able to discover my heaven so I am not a heavenly material. If I will be placed in heaven, then I will turn heaven into hell.

I heard a story about a rescue team visiting a flood affected area and one of the team members saw a scorpion.

He replied the scorpion is doing what he must do, so I am doing what I must do.

This is one self-aware, self-purified Supreme Being.

Every human has the ability to be self-aware and do self-purification to live like a supreme being.

A supreme being does its best to others expecting the worse from others, if he or she gets what was expected, Supreme Being is being rewarded and will get elevated to a higher level.

If the others do nothing in return, Supreme Beings feel ignored, but will be OK with it.

If someone wants to do something good in return, that person will be requested to do the good deed to some needy person.

Whatever faith you choose, make sure to be self-aware and do self-purification first.

You will be in a state of worship 24/7 and feel the taste of heaven in this world.

Even in your dreams you will be doing the righteous deeds.

Today I was told it is the end of the world. This is true, it could be a beginning of a new beautiful world if we all decide to become self-aware and decide to self-purify our self and turn Harrisburg into Heavenly Harrisburg, where love & happiness tolerate and justice prevails. All the other cities in the world can duplicate the process.

I must transform myself into heavenly material.

With self-awareness and self-purification we will be many bodies and one heart. I thank you for giving me the award I know I do not deserve, I will accept it conditionally. If I cannot prove to myself, or you feel later that you made a mistake, I will be more than happy to return it or give it to a deserving candidate. I know that you are OK, I am not OK, and that is OK.

PEACE

Healing Tip 173

There are two kinds of energies, potential (or hidden) energy and kinetic (functional) energy.

To lift an object you use energy and that object will have potential energy.

To maintain it, you need energy, otherwise the gravity will overcome and that object will drop and potential energy will convert to kinetic energy (functional energy).

Same way when I work on myself I gain potential energy depending upon the amount of work.

To keep myself at that level, I also need maintenance energy.

Depending upon the amount of energy will determine my ability to overcome any obstacle or challenges that come my way and still have enough energy to elevate myself until I reach to the top of my desired destination.

If I am arrogant, judgmental and revengeful, these traits will consume so much energy that I will never be able to live to my full potential.

Be humble, nonjudgmental and forgiving, and conserve your energy to reach your goal.

Healing Tip 174

Buddha said something like this - Life is suffering.

To me, life is a beautiful journey offering me multiple opportunities.

However, there is always suffering in life, therefore life is always suffering.

Healing Tip 175

I must quit worrying about the future, I do not have control over others, and I must control my actions. I must stop worrying about things, I have no control over, I will only worry about myself, and I must have control over my mind and body, if I will have control over my body and mind, it will be easy to control my emotions. I must control my body and emotions.

Healing Tip 176

You have another three choices:
Blame others and be bitter,
Blame myself and be miserable,
or
Learn and grow.

Healing Tip 177

You must not only be able to live with someone, but without them as well.

Healing Tip 178

I heard the story of 3 men who ended up on a deserted island.

After many years, they found a magic lamp and a Genie appeared.

"I will fulfill one wish for everyone," she said.

The first man said "I want to go home" - and his wish came true.

The second man asked for the same and his wish came true.

Then, the genie asked the third one about his wish.

He said "It is lonely without those two, I want them back here."

There 2 main lessons to this story:

Moral #1 - Pay attention to the company you keep.

Moral #2 – Communicate with those around you

Healing Tip 179

I hear all the time that life can be worse, or life can be better. The quality of your life is all based on your frame of mind. I usually hear three responses to: "How are you?"
Same old same old
Life can be better
Life can be worse
All three can be true, but if you think positively, more good will come.

Healing Tip 180

Stick up for yourself, and do not let other people take advantage of your kindness.

Healing Tip 181

Special thanks to Kristen Barbacci from Harrisburg Pa for being the inspiration for this Healing Tip:

The present moment is the best present (gift), and is defined by the choices we make.

The most difficult choice for me is what to speak, when to speak, how much to speak and whom to speak or keep my mouth shut.

Healing Tip 182

I came across a person in the psych ward many years ago.

He had schizophrenia, and was extremely anxious and hard to treat.

The medications used turned him into a dysfunctional zombie, and cutting down the medications resulted in an anxious, violent personality.

There was no happy medium.

He was considered a hopeless case.

Many years later, I was pleasantly surprised to see that person as my patient, whom I did not recognize at first.

He was married and had his own place and a job.

I asked him about his transformation.

His story was very inspirational. He said "I was tired of living, I just wanted to die, my every attempt failed and no one helped me to kill myself or live a healthy life."

"I was forced to look within me. I experimented with different things, and I discovered that if I spent three hours alone in a dark room and prayed, the rest of the 21 hours I lived anxiety free."

How do you pray? - I asked.

"I loudly say help me Lord nonstop for 3 hours."

"I cannot afford to miss my prayers and medications because I know I will drift down to my old self."

I was stunned. He was more self-aware than I was.

I have not acquired that balance in my life.

I thanked him for being such an inspiration.

Healing Tip 183

HAPPY FATHER'S DAY!

On Father's Day I look back and remember my father, and his impact on me.

I am so fortunate to have a father like him, and I did not realize that when he was around.

I will now show a great story from my life.

When I was in fifth grade, my desk-mate bought a new book bag.

I had an old book bag which I was happy with, but everyone else made fun of it.

I went to my dad and asked him to buy the same book bag.

He asked me why I want that particular bag. When I told him my reason he told a story as usual.

A deer was drinking water in a stream, and he saw his shadow in the water.

He felt so happy to see his beautiful horns, but when he looked at his skinny legs, he was very unhappy.

Suddenly he heard the lion roar, and he ran. His skinny legs saved his life until his beautiful horns got stuck in bushes and the lion got him.

I knew what my father was telling me. But I wanted the new bag.

He told me that he could not afford it this month, but he would buy it next month.

I was happy, and I told everyone in the class that I would have the same bag.

One day we were playing soccer and left our bags on the table, and when we came back that boy's bag was missing.

He reported to the principal that I had stolen that bag because I wanted it.

It was a painful experience.

Later on that boy came and apologized. He said "I know that you did not steal that bag, I just did not want to get in trouble with my parents."

I just felt sorry for that boy instead of getting angry, because I remembered my dad's story.

A few days later my dad gave me the money to buy the new bag, I told him I did not want it anymore.

He said: "I recommend that you should give this money to someone who needs it more than you do, and you will be much happier than having a new bag." I did that and I truly was very happy.

I kept that old bag until I graduated from High School. I did not care what anybody said about it.

Now I know why I drive a 1994 Van and a 1998 Honda in spite of all the criticism by a lot of people.

It was not about the bag. It was principal I learned to live by. THANKS DAD. (I know you hear me, and are smiling with me).

Healing Tip 184

Since childhood I was told that forgiving others is a good thing.

I found it very hard and frustrating, and it never turned into a good thing for me.

Later in my life I was told to not only to forgive, but also to forget.

I found it even harder to practice, as I felt I always kept going in circles and ended up frustrated and angry. No one was happy with me, and I felt like I was going nowhere.

I learned to forgive, learn, memorize and earn my credits and grow so that I did not have to go through the same process ever again.

Now life is great, I am meeting new people, learning, earning my credits and growing, soon I will earn my degree - FD (Doctor of Forgiveness).

So I can send gifts to everyone who helped me earn my much needed degree.

Healing Tip 185

Special thanks to Lisa Davis for being the inspiration for this Healing Tip!

You are your best judge! You have witness everything that has ever happened to you. Be brutally honest with yourself, and give a just verdict. If you do that, you will serve your time well and live a peaceful, crime-free life.

If you do not give fair judgment, you will stay captive of yourself, and sooner or later someone else will pass the judgment for you. You may not like it, it may not be fair, but you will still be forced to serve your term.

Judge yourself first.

Healing Tip 186

Special thanks to Shelly from Harrisburg, Pa for being the inspiration for this Healing Tip!

If I discover my evil, I will see my evil; I will know my evil, I will tame my evil, and I will cage my evil. I know how to handle evil.

Then I will do no evil, see no evil, hear no evil or talk no evil.

My action and reaction will be healthy.

I will be an evil proof human, and live a peaceful life.

If you tame your evils you win, and if your evils tame you, then you lose.

Healing Tip 187

Sometimes you work very hard to help others but then realize that they never wanted or needed your help.

It is like teaching a pig how to sing. If you think you're frustrated, ask the pig!

But if you love the pig, you can still attempt to teach him, as long as her can pick the song.

Healing Tip 188

A tree is known by the fruit it bears.

If I have peace with myself, then I will be a peace tree, and a peace maker.

If I have trouble with myself, then I will be a trouble tree and be a trouble maker.

Grow into a peace tree, and share your fruit with the world.

Healing Tip 189

Special thanks to Glenmarie Costa for being the inspiration to write this Healing Tip!

Life offers us a constant stream of choices, the next choice depending upon the previous choice.

Each choice is attached with reward or consequences depending upon the choice.

Usually one unhealthy choice is followed by another unhealthy choice, unless you learn from the unhealthy choice and never choose it again.

One example is choosing what to eat. There are a lot of resources for popular consumption, like McDonalds, Burger King, Wendy's or Subway – each of which claims their food is the best.

None of them can be as healthy as when you make your own food from scratch using fresh and healthy ingredients. Of course it is a lot of work, but it is worth it in the long run.

Healthy Choices = Healthy Pure Heart = Happy Purposeful Life.

Unhealthy Choices = Unhealthy Heart = Unhappy Miserable Life.

Healing Tip 190

I used to have many high expectations for others, and the majority of them did not live up to my expectations.

This disappointed me, as I always believed something was not right with the other people.

I kind of gave up on them.

To my surprise, I realized that I cannot live up to my expectations.

I discovered that it was always I who was the problem. I saw my reflections in others, and blamed them instead of thanking them.

Healing Tip 191

A man ordered burnt toast, half spilled coffee, and a tasteless omelet with no salt and pepper.

The waitress was surprised and said that she never had this kind of order.

"That is the way my wife makes my breakfast, and her cooking is the only thing that keeps me healthy and well. I love her" - the man replied.

Healing Tip 192

Religion + Healthy Understanding = Righteousness = Self Peace + Peacemaker.

Religion + Unhealthy Understanding = Unaware Human = No Self-Peace + Weapon of Mass Destruction.

The unaware human lies to themselves and they do not knot it. They are the ones who need the most help, as they not only live in their own misery, but they create misery for others as a way to solve their problems. These people can only be treated with understanding, love, and kindness to help them become aware. Your love should be more than the hate they harbor to win their heart.

Healing Tip 193

What I see depends upon the vision I have.
Mother Teresa saw darkness in herself.
George Bush and Osama Bin Laden saw darkness in others.
All three are religious people to some extent.
Any religion is a gold mine.
Mother Teresa picked up the gold, worked on it and turned it into pure gold, but still she felt she could be better.

The other two picked up stones and turned into religious nuts.

Healing Tip 194

If you love yourself, then you know what love is. Only then are you able to truly love others who are near and dear to your heart.

Healing Tip 195

My quality of life depends on many variables and my attachment to them.

In the uterus, I was dependent on the umbilical cord - that was my life line.

Since my life line was cut, I am my own.

The attachments you develop with others are either healthy or unhealthy.

Each attachment costs mental, physical and emotional health and the one those are not cost effective must go.

The less attachments you have, the greater quality of life

The less attachments you have, the greater quality of life you will receive.

Healing Tip 196

With self-discovery I will appreciate everyone and everything that was aiming to help me intentionally or unintentionally or aiming to hurt me intentionally or unintentionally and end up helping me to become self-aware.

Start by trying to understand everyone and everything.

Healing Tip 197

Humans are the supreme creation, best of the best, with the ability to be the worst of the worst.

I used to share my opinion with others, hoping they would benefit from it, to my surprise when I told them the greatness I saw, they not only did not believe me, but they asked me what I wanted from them.

When I mentioned any shortcoming to anyone, I was demanded not only an apology, I was asked to praise them as I did to others.

This cost me a lot emotionally, mentally, financially and physically.

I could not afford to operate that way.

My new mission statement is:

Each human is a masterpiece.

I must value everyone's opinion about me and learn from everyone.

As far as my opinion is concerned, I will share only upon request.

In most cases, it will be by appointment only and advance payment is expected.

I am well off and live a happy life and most of the people are like me. (That's what they tell me).

Healing Tip 198

Special thanks to my dad who was a great father and teacher.
When I was young I was restless and self-destructive.
My father loved me, was caring and had a lot of patience and never gave up on me.
With his hard work, I became much less restless and self-destructive.
He was a teacher, and he must have had that secret recipe, because a lot of his students shared the same opinion about him.
At the end of his life he was happy and rested in peace before he went to his final resting place.
Now I am mature, restful and productive like my father.
If I will be able to help the younger generation, not to be self-destructive and stay healthy. Like my father, my last days will be happy days and I will rest in peace before I will rest in peace in my final resting place.
Otherwise there may be no peace here to rest in.

Healing Tip 199

Humans are supreme - best of the best with a lot of potential and unlimited possibilities, but they can be captives of fear.
The fear if I live up to my full potential is that people will not like me. I will be an odd ball, not able to enjoy the gossip and not miserable and will not need company to share my miseries with others and no one will understand why I am always happy.
I have seen greatness in each human I ever had encountered.

Even with sick people or prisoners, I can relate and feel their pain and suffering. But I know that with a healthy frame of mind these people have the potential to be best of the best, only if they take responsibility for their actions and learn and discover the Supreme Being that is waiting to emerge.

Iron needs to go through an oven to turn into steel. Humans need pain and suffering for self-purification.

Conquer the fear and live a supreme life.

It is OK to be a happy odd ball

This is the End of the Trail of Healing 202 - Lesson #2

If you have successfully climbed the trail, precede to Lesson #3 - Healing Tip #200

GOOD LUCK

Healing Tip 200

Human behavior is a combination of learned behavior, emotions, and reasoning.

Learned behavior is mostly primitive behavior, and most people respond to those impulses.

Emotional behavior lacks reasoning.

Reasoning is controlled by the cortex - the higher center of the brain.

Healthy people are predominantly controlled by the cortex and keep a great balance of the other two factors and live very healthy and balanced life.

Cortex driven personality is a trait that is hard to acquire and difficult to maintain, but the reward is a great quality and quantity of life.

Healing Tip 201

I always felt that I am an odd ball with a lot of rough edges, and most people made me aware of that.

I always tried very hard to fit in but always ended up being unhappy.

Now I do what I have to do and live a happy and successful life.

I may not win the popularity contest, but that is OK.

Work on yourself, not others, and the rest is history.

Healing Tip 202

Everything I see depends on my vision.

It is not the book; it is I who reads the book.

Same way I see others and other things in a different way than they or those things are.

I relate to others and other things depending on my vision. (Healthy or Unhealthy)

With healthy vision, I will see my reflection in everyone and everything, and resources for me to grow. There a story, a long time ago a boy was crazy about a girl. The king heard that story and wanted to see the girl.

When he saw her, he said "I do not know what he sees in you."

"You are not even pretty."

The girl replied, "You just do not have his vision".

Healing Tip 203

You must love and enjoy everything and everyone.

You must not become dependent or attached to anything or anyone.

Like standing on a beach, waves come and kiss your feet, enjoy it, but it goes away, enjoy the breeze, enjoy the view, there is always plenty to enjoy if a wave comes back.

P.S. Love is a healthy attachment and dependence is unhealthy.

Addendum: Sickness, pain, bad things are not really bad, just reminders. The quicker I learn my lessons, the quicker they will leave me.

Healing Tip 204

Humans are nothing but mirrors, they show my reflection. If you are an honest mirror with a healthy frame of mind, then you will love every human for their help and be a wise mirror, shown to anyone on written notarized request only and at their own risk.

Healing Tip 205

Life is full of contrast.

You cannot appreciate a sound healthy mind if you do not know what a fragmented mind is like.

In this same way, you cannot appreciate a sound physical, emotional, spiritual wellbeing or any other thing if you have not known the opposite.

One person said - "Only if I have a peaceful mind for one day, only for one day, I will give up everything I have, just to have one peaceful day. I had a peaceful mind, I lost that peace, and I wish I would have appreciated, valued and kept it. I wish I can turn back the clock. But I cannot."

Encounters like these make me realize how fortunate humans are, so gifted.

Only if they appreciate, value, enjoy and keep it while they still have it.

Healing Tip 206

Life is simple and straightforward.
I did a great job and made it very complicated.
Work on setting your priorities straight.

Healing Tip 207

You will accomplish goals you never dreamed possible.

Healing Tip 208

MY NEW AND IMPROVED COMMANDMENTS!

I must constantly gain useful knowledge; it is a form of worship.

Difficulties in life are gifts, so I can understand the purpose of my life.

I must learn from history and from others, so I do not make those mistakes.

The secret of life is whatever I am, wherever I am, whatever condition I am in, I must be happy and content.

I must earn an honest living, and take care of my body with healthy food, exercise, good sleep and feed my mind healthy thoughts.

I must love myself conditionally and everyone and everything unconditionally.

P.S. These commandments are not engraved in stone and still need a lot of improvement.

Healing Tip 209

What is the difference between a good cook and a good doctor?

A good cook cooks delicious food for others, so he or she can afford to cook healthy food for him-self or herself to keep cooking for a long time.

A good doctor takes care of his or her health first, so he or she can provide good healthcare to others for a long time.

What is common between a good cook and a good doctor?

A healthy waistline.

Healing Tip 210

Anyone who likes to give free advice to others should remember that it may cost a lot.

One may end up friendless, spouse-less, peace-less, family-less and may be everything-less.

You must question myself before giving free advice, how much it will cost and if you can really afford it.

Healing Tip 211

Humans are treasures, especially the ones who make you very uncomfortable.

Those are the ones who are the most challenging and most rewarding.

Several years ago, one of my patients on his first visit said - "I know you will dump me. Everyone dumps me. I know I am not an easy person. I know, I live with myself. I am trying to learn to put up with myself. I do not have a choice. Other people do, they just blame me and walk away. I will not blame you if you do the same."

He continued to talk for the whole session.

I said at the end of the session - "Let's try, time will tell - try to be positive."

It was a very difficult several months; he talked all the time, contradicting himself at times. I felt his struggle, but felt helpless. He did not keep the diary as I told him to do, as a matter of fact; he never did what I told him to do.

I was hoping that he would give up on me because I felt guilty telling him the same things.

To my surprise, he started to do well, and was able to do what I never thought he would accomplish.

He thanked me for all my help, and said - "You are the first one who really helped him!"

Healing Tip 212

Special thanks to Naeem Sheikh from NJ who inspired me to write this Healing Tip!

A man saw an angel working on his lap top.

"What are you doing?" - The man asked.

"I am updating the list of people who love and serve GOD the most." - The Angel replied.

"Is my name in that list?" the man asked.

"No, but I see you're on the list of people who take care of themselves so they are able to love and serve the creation, and also your name is at the top of the list of the list of people loved by GOD the most." - The angel replied.

Healing Tip 213

I have two open ears to remind me to keep an open mind.

I have two eyes with convertible covers to see pure and healthy things.

I have a closed mouth, which is port of entry for healthy food and emergency outlet to call for help.

Using the modern technology, like fax, internet, texting and other ways of communication are more effective ways to communicate, as they can be tracked back and corrected, which is not possible with obsolete verbal communication.

Teaching healthy verbal communication is hard work.

Healing Tip 214

One day I retraced my footsteps back to childhood.

I noticed I always went astray and was getting into trouble until I submitted to GOD.

Since then I was able to stay on a straight path.

I felt good, and thought that I was better than many people and thanked GOD.

GOD answered - "Oh, my beloved creation, since you submitted to me, it is me who is carrying you to a straight path, and those are my footsteps."

"I have made you weak and you are not able to deal with distractions and intoxication around you to stay on the right path."

"So be humble, never look down to my creation, and serve my beloved creation unconditionally, and I will continue to carry you to the straight path."

Healing Tip 215

This universe is full of natural resources.

Gold mines, diamond mines, crude oil - a few of countless resources.

All natural resources need purification to make it most useful including humans.

Healing Tip 216

Humans are captive of their desires and dreams.

I was the captive of my desires and dreams.

My life was nothing but a struggle, and still I always had unfulfilled desire, and was always running after my dreams.

Finally I put my desires and dreams on the back burner, and start working on myself. I was able to love myself, it became easier to serve and learn from others.

I was surprised that not only my desires were fulfilled, but also all my dreams came true.

Healing Tip 217

Special thanks to Sarah Siddique for being the inspiration to write this Healing Tip!

The soul has a concealing seed of compassion.

Crushing toxicity breaks the shell, and the seed of compassion is released.

The seed of compassion needs a land of misery, fertilizer of suffering, and water from the river of pain to grow and blossom.

A purer and stronger soul emerges, waiting to repeat the cycle, until all the misery, suffering, and pain is consumed.

Healing Tip 218

I receive constant reminders and rewards for my actions and reactions.

Any set back is a reminder, and any pay check or victory is a reward. They are both equally important either to modify or keep doing what I am doing.

The more you take responsibility for you actions and reactions, the less and less setbacks and more and more rewards will follow.

Healing Tip 219

You must practice what you know, and keep your book with your to-do list and not -to- do list handy and live by it.

Healing Tip 220

Special thanks to CO Marlin Frank Houser from Camp Hill Prison (I call it Paradise) for being the inspiration for this Healing Tip.

There are three kinds of people all serving their unique purpose.

The first kind knows their purpose and serves their purpose well.

The second kind provides the purpose to the first kind and serves their purpose.

The third kind have mixed traits of both.

George Bush and Osama Bin Laden served their purpose so did Mother Teresa and Florence Nightingale.

Healing Tip 221

Each human is blessed with unique gifts.

Depending on how those gifts are used, privileges are granted, taken away or kept.

Random testing is done for quality purposes in the form of setbacks, difficult situations or difficult people to test if an individual may keep the privileges, additional privileges may be granted, or privileges taken away.

Human's potentials are limitless regardless of how they are being utilized.

Healing Tip 222

Some feel that we live in a dysfunctional world.
So we have to be dysfunctional to be functional.
I feel I am dysfunctional in some areas and to live in this world I must be functional.
Dysfunction I see outside of me is my reflection and reminder.
If I continue to learn and work on myself, then I will become functional and grateful.
P.S. For every problem there is a solution and for every disease there is a cure waiting to be discovered. If you cannot find it I are you looking hard enough?

Healing Tip 223

Work is as sacred as worship.

Healing Tip 224

Special thanks to Melisa Young of Hamilton Health Center for being the inspiration to write this Healing Tip.
The people who have given you trouble and ended up helping you can be categorized into three groups:
People you must apologize and thank,
People you must thank,
And People who you owe an apology.
All three groups have equally helped you grow.

Healing Tip 225

Knowledge without application is not practical knowledge.

I know that humans are born with a primitive functional brain (Limbic System) which continues to evolve under the influence of nurturing.

By age 25, humans have the foundation, the product that is mostly limbic system driven, influenced by the nurturing it received.

By age 25, the cortex (the higher reasoning center) becomes functional, reminding humans about the defective foundation and its consequences and keeps reminding me to pay attention to natural talents and use it to live up to my full potential.

This is not an easy process, because the 25 years' strong foundation is not easy to dismantle or rebuild.

Only (and only) if I become a cortex driven person and everything I do must have a healthy reason I will accomplish my goals.

Now if I analyze myself, I was born in a home, to a family and I had no control.

The nurturing I received I had no control of. I was primitive brain (Limbic system) driven.

All those were not my choices. By age 25, I was a product mostly of outer influences and had a complex foundation with a lot of learned behaviors.

By age 25, my cortex (Higher center of the brain) becomes functional.

I know I wanted to do well so I worked hard but did not question my learned behaviors (My foundation) as I should have.

I built a future by thinking, profession, family, money will make me happy.

All this was built on an unhealthy foundation and was weak and defective.

The more I tried to be happy, I always felt something was missing.

It was like trying to fix a house top down; it just did not work.

Finally I questioned everything I had learned, and reasoned it.

It was hard work, dismantling the foundation and building a new me, but I was able to have a strong healthy foundation.

Now I live a happy life.

To fix anything, fix the root cause and everything will fall in place.

Healing Tip 226

The whole universe, including my mind and body, follows the laws of nature.

If I choose to do otherwise, then I am not only creating trouble in the universe, but even my mind and body rebel against me, and I will have no peace with myself and others.

If I continue, first I will lose quality of my life followed by quantity of my life, and die unnaturally.

Healing Tip 227

Special thanks to Mr. B. Herb from Camp Hill Paradise for being the inspiration for this Healing Tip.

This world is a stage and we all are the actors on this stage.

My success and failure is determined how well I know my role, and how well I play my role.

My job is not to question others, but to accept everyone and learn from them so I can perform my role well and earn my Oscar award.

Addendum: Everyone will have his or her awards and rewards depending on how well he or she understood and performed his or her purpose. It is a simple flip of the coin - Judge you lose, Learn you win.

Healing Tip 228

Special thanks to my anonymous friend for being the inspiration to write this Healing Tip.

Someone said "Stupid people drive me crazy." What I have noticed is that there are two kinds of people.

People with a low IQ and people with high IQ.

Success and happiness only depends upon awareness.

Unaware low IQ people do things that may upset others and drive some people crazy.

Self-aware people with a low IQ do not upset others.

Unaware people with a High IQ are likely to be driven crazy not only by unaware low IQ but also by unaware High IQ people as well - they drive themselves crazy.

An aware person with a high IQ knows driving anyone or driving themselves crazy is a serious crime against the LAW OF NATURE and is punishable by losing the most important privileges, quality and quantity of life depending on the severity of the nature of the crime.

Healing Tip 229

The human mind and body live together like husband and wife since birth.

The body matures faster than mind.

By age 18, the body is fully matured with an immature mind.

Approximately by age 25, the mind is mature and becomes functional.

How well your mature mind and body work together determines the quality of your life.

If the mind is hyperactive and the body is underactive or vise verse, it is a bad marriage.

If both work together and act mature, it will be a honeymoon.

Healing Tip 230

In humans there are three main important components.

Body, Mind and Heart.

If I have a healthy mind, body and heart 100% compatible with each other and they work as one unit, then I have a perfect life.

A wondering mind cannot focus 100% on the task and cannot perform 100%.

Very soon, I will be a wondering heart and unhealthy body.

My wandering mind will seek smoke, alcohol and drugs to stay focused and for instant gratification, affecting heart and mind adversely in the process.

Wandering heart will seek pleasure with multiple sex partners, overeating and other unhealthy things damaging mind and body.

The same way the unhealthy body will not be able to perform needed functions properly.

This chaotic state will consume the quality then the quantity of life, and unable to live to full potential and serve a lesson to others.

Healing Tip 231

Everyone and everything is there so I can learn and grow.

If you ponder and learn, then you will love the outcome, and love everyone and everything.

It will be a WIN-WIN situation.

If you judge anyone or anything, then you will become captive of love and hate.

A Lose-Lose situation.

Pondering, learning and loving make you a free person and you will have a happy life.

Judging and disliking make you captive, and you will live an unhappy life.

Healing Tip 232

Special thanks to Patricia Botti, RN from SCI Camp Hill Paradise, Camp Hill Pa for being the inspiration to me for writing this Healing Tip.

Every medical meeting I go to, everyone is talking about the bottom line.

I know simple math. $A - B = C$

A) The top line represents my assets. (You are your most precious asset)

B) The middle represents my expenses including your maintenance.

C) Bottom line.

Everyone and everyone is trying to cut costs to have a better bottom line; instead of investing into the top line.

Investing on yourself is the secret to have a great bottom line.

Healing Tip 233

Laughter is superficial; I must discover real joy and inner happiness.

Healing Tip 234

A well-functioning computer must have a safety system in place to function well.

A well-functioning security system not only blocks all viruses and harmful things to enter, but it also keeps scanning for any harmful and undesirable elements within the computer and destroys and eliminates it.

It is a constant process and that safety system needs constant updating to meet the new challenges that may arise.

Healing Tip 235

There is an old saying: "A book cannot be judged by the cover."

Books are not to be judged, cover or contents.

You must learn from books especially the ones you do not like or cannot understand.

All you can do is know yourself, know your cover, know your contents, and keep learning and be able to come to the conclusion that you do not even know a fraction of yourself.

Healing Tip 236

Special Thanks to a young woman for being the inspiration to write this Healing Tip!

Mother Nature does the nurturing of everyone and everything.

Whatever (ever) happened to me until today was planned by Mother Nature. After each event, I always was at the cross road, to choose from a path to lead to the next cross road.

If I ponder without questioning, then I will discover the hidden wisdom in every minute detail of my life.

I will know my purpose, and wherever Mother Nature places me at the cross road, I will make the right choice.

Healing Tip 237

Know that you are a good person. No one has the right to tell you differently. You do mess up at times, and you must apologize and work on yourself.

Healing Tip 238

"I have a hard time respecting someone who disrespects me."

I humbly and respectfully disagree with this statement.

When I review my life, every time I disrespected anyone or anyone disrespected me, they were missed opportunities for self-growth.

Whenever I feel disrespected or I have no respect for others, then I must do my soul searching and be brutally honest with myself. I will discover that I have not earned the respect I deserve.

Disrespecting others is a sign of unawareness.

Feeling disrespected is just a reminder, and you must have very high regard and respect for those people for making me aware of my shortcomings.

Healing Tip 239

NO PAIN NO GAIN.

To gain self-awareness I have to go through pain, but that will be most rewarding part in the long run.

With my eyes open I can see the world outside of me, and my outer life depends how I interact with the outside world.

With my eyes closed, I must be able to see the world within.

That will take a lot of hard painful work.

You will discover the universe within, full of resources, treasures and potentials.

Healing Tip 240

A long time ago a wise guy said to a wise person - "There are only two scholars in the world, you and me, but I am not too sure of you."

"I can teach you to be a scholar, but of course you will never be as great as I am."

The wise person replied - "I can learn from you, you sure are a treasure."

That wise person not only learned from him, but also learned from everyone and everything and she was a great scholar of her time.

Wise people not only coexist with everything and everyone peacefully, but they learn from everyone and everything and are grateful to everyone and thankful for everything.

Healing Tip 241

Sometimes people do not fit together, like two blades attempting to fit into a sword-holder. It just doesn't work, so stop trying to force them together.

Healing Tip 242

I must keep pondering, and keep evolving to be a better person. Humans change, but stones do not change.

Healing Tip 243

If I can relate to others, then I can learn so much.

One of those outstanding encounters was with a young patient who lost his eyesight in an accident.

When I visited him I tried to comfort him.

He smiled and said "Thank you for your concern, I am more than fine.

I was blind, now I have acquired vision.

I can see within and what I have discovered, I am so thankful.

True I cannot see but I sure do have a vision."

I was speechless.

Someday I wish to acquire the vision he was talking about.

Healing Tip 244

There is not substitute to human touch.

Healing Tip 245

Special thanks again to Patricia Botti RN from Camp Hill Paradise (Formerly called Camp Hill Prison) for being the catalyst to write this Healing Tip.

Each human is a physical being as well as a spiritual being.

A healthy balance turns a human being into a supreme being.

An unhealthy balance makes anyone anything but a Supreme Being.

Physical health needs regular exercise, healthy diet, and a good night sleep. Healthy spiritual health needs constant soul searching and continuous self-purification.

Healing Tip 246

SPECIAL THANKS TO MY MOM AND DAD WHO ARE IN HEAVEN FOR LIVING EXEMPLARY LIVES— MY REAL HEROES!

ALSO, A SPECIAL THANKS TO MY TWO SISTERS REVERAND CYNTHIA MARA AND SHAHIDA BEGUM, AND TWO OF MY PATIENTS RENEE STICKLE AND MILDRED BOROTA WHO HAD SERVED WELL AND ARE STILL SERVING, MY HEROES AND I KNOW THEY WILL GRADUATE LIFE WITH FLYING COLORS.

~~ HAPPY END OF THE YEAR TO ALL OF YOU!! ~~

Back home the school year is March through December.

December to celebrate if I pass the grade, enjoy two months and be ready for the next grade. If someone fails, they go to school January and February and take their exam. If they fail again, they had to repeat the same grade.

I learned early in life to work hard and pass, knowing that next grade would be harder, but my goal was to graduate.

When I came to the US, I could not quite understand celebrating New Year, making resolutions, only to repeat the process next year.

With the end in mind, focusing on the goal always gave me purpose, it kept me going instead of getting lazy procrastinating.

We all are serving a life term here, if I serve well, then I will graduate and go back home.

Healing Tip 247

Eating healthy is not only is the LAW OF NATURE, it is the foundation of good health.

It is like using the high quality of gas in a car: the engine will work fine, and it will cause less pollution.

Healthy food will not only keep me healthy, but it will add the least amount of smell in the air.

I may have to compromise and ignore or fool my taste buds, but it is worth it in the long run.

My ideal diet is 50% green's, 33% fat and protein, 14% High Fiber, carbohydrates and unlimited water.

This is the ideal diet to stick to.

Healing Tip 248

I was unaware.
My mind was a battlefield.
My life was a battlefield.
My universe was a battlefield.
NOW I AM AWARE.
My mind is peaceful.
My life is peaceful.
My universe is peaceful.
Now I know my mind was fragmented and shattered.
It was piece-full (Pieces with rough and sharp edges).

With awareness, I was able to place all pieces in the proper places.

Awareness is the key to turn hell into heaven, from piece-full to peaceful universe.

Healing Tip 249

I have discovered to treat patients effectively. Unconditional love is a prerequisite to learn their history and come up with an effective treatment plan.

When patients sense my love, they open up and tell me every detail I need to know to come up with the correct diagnosis.

I also have discovered that I am the most difficult patient I ever dealt with.

It is so hard to love myself unconditionally, when I know what I have done and what I am doing.

When I started loving myself, I was surprised I discovered diseases in me I never knew existed.

I also noted that I listened to my walk more than my talk.

The more I love myself the more I am learning from myself and coming up with correct diagnoses and treating myself more effectively.

This is an on-going process. My problem list is constantly changing.

Treating myself effectively is helping me tremendously to treat others.

LOVE, LEARN & SERVE is universal antidote to any ailment.

Healing Tip 250

Special thanks to Shelly Young for being the inspiration to me for writing this Healing Tip!

This universe is a perfect human lab, run by all mighty with perfect wisdom with perfect checks and balances.

JUDGE NOT & I WILL NOT BE JUDGED.

True, but I must learn from everything and everyone and keep judging myself, knowing that one day I will be judged.

I must not only know myself, but I must keep digging in my past and use all the resources at my disposal to gather all possible information and evidences and information about me.

I will discover a gold mine (gold mixed with dirt, stones and other impurities).

It is a hard job, but is the most rewarding job.

I will discover people who were unjust and did bad things to me, abused me.

I also must remember what I did to them.

I know I must not judge. I must have a valid reason for those events.

If I know almighty is running the show, and those events may be a reminder to me to get my act together, or to learn must needed lessons so I can relate to other people who went through similar experiences or worse experiences.

If that is the case, not only must I forgive all those people, but pray for their forgiveness and send them thank you notes for their contribution to my growth.

This is the easy part - now the hard part.

What about those people I was unfair to and did bad things to them?

I must come up some way to compensate for their suffering if still possible; otherwise I must do a lot more good deeds to neutralize the toxicity of my crimes.

I must become the strongest advocate against the crimes committed to me and I committed to others.

I must learn to love, learn and serve myself then others and everything unconditionally.

I must do my homework honestly and be able to seek self-forgiveness and hope when I will be judged, I will be forgiven.

This must be done before the deadline strikes without warning.

The earlier I do it, the sooner I live peacefully with myself and everyone and everything and hopefully my life stay the same after death.

I am the judge, I am my advocate and I am on the stand, I am the advocate against me, my verdict may determine my fate.

Healing Tip 251

Two boys met outside and the first son said to the other - My dad is so stupid.

The second son said - My dad is the most stupid dad.

The first son said his father wanted him to call into his office and say he did not want to work today. The telephone was there in the room he could have called himself - but he could not think.

The second son said "he gave me money to buy him a car, but forgot to tell me which kind and the color he wanted."

If you do not treat those superior to you well, then they might not treat you well either.

Healing Tip 252

Special thanks to the inmate for being the inspiration to me for writing this Healing Tip!

I met an angry inmate in the hole.

I asked him why he is so angry.

He replied – "I am in the hole for no fault of mine, and the guards are mean and unjust.

I am going crazy in here with no TV, no yard, no one to talk to, and the days are so long it is a hell." Then he showed the verse from the Bible - The Innocent people will be prosecuted and the criminal will be free.

"I am innocent and I am in the hole - that is not fair."

I said: "You are not the first one - Joseph was innocent and ended up in prison, but he believed in the higher power, so LORD was in his Cell.

He was never alone. Who could be better company than the LORD?

In your case, it seems that other people cannot put up with you, and it seems you have trouble putting up with yourself."

He got really angry and said – "You can talk but you never have been in the hole - what the hell do you know?"

I replied – "On the contrary, I spend three months in isolation, alone in a worse cell than this one. If you have time I can tell you my story."

"I have all the time in the world - Go Ahead."

When I graduated from pharmacy school I was 22 years old and I applied for a job and went through a health screening.

I was told that I have 3 months to live because of extensive TB which was not worth treating, and they recommended for me to be in isolation until I died.

I spent three months in isolation waiting to die.

Those three months were the most rewarding of my life.

I discovered my eternal mate, my sustainer, the master planner THE LORD. I found out I am not alone.

That was a reminder from the LORD, not a punishment.

During that time, I discovered I was not innocent, I discovered my sins and decided to do more good deeds to neutralize the toxicity of my crimes.

Later I discovered that the X-Ray technician placed a wrong name plate on my X-Ray and I was perfectly healthy.

Since then, the LORD has never left me, in spite I left the LORD many times but whenever I turned, I found LORD waiting for me.

The inmate thanked me for sharing my story.

Healing Tip 253

ONE LESSON

Dad asked his son "Son, why do you do everything wrong?" The son replied, "It is easy dad, I can teach you too."

Moral: Missed opportunities to understand each other and grow together.

Healing Tip 254

This life is a dream and I am the main character of my dream.

If I live my dream to its fullest, then I will have not only the heavenly ending but also beginning of heavenly eternal life.

Healing Tip 255

Special thanks to my patients (My Heroes) for being the inspiration for this Healing Tip!

I usually ask this question to my patients - WHAT IS THE THING OR PERSON THAT BOTHERS YOU THE MOST.

The responses amaze me and I have learned so much from those responses.

I chose to ask myself the same questions, and I realized that I am the main reason for anything or anyone bothering me.

By doing research, I discovered that I was the one bothering them a lot more than they bothered me.

I decide to declare a peace treaty with myself that I must learn from everyone and everything, be thankful to everyone for everything and never act or react in a way that bothers me in any way.

DO YOU KNOW WHAT ARE THE THINGS OR PEOPLE THAT BOTHER YOU? Knowing the answer will determine the quality of your life.

Healing Tip 256

You may be struggling with three basic terms and conditions:
Healthy Diet
Regular Exercise.
7 to 8 hours of a good night's sleep.

Every day brings new excitements, challenges and happiness of its own kind and makes every day more perfect than the day before.

Healing Tip 257

I see with my eyes outside of me.
I see with my heart inside of me.
Having a blind heart is worse than blind eyes.

Healing Tip 258

With awareness I will see life like this:

L - Stands for LOVE myself, learn about myself, and keep searching for truth and become self-aware.

I - Stands for INSIGHT of my shortcomings and continue to strive to better myself.

F - Stands for FOCUS on myself learning from myself and others always FOCUS on the end.

E - Stands for EXCELLENCE in everything I do, have a pure heart and peaceful mind.

I do have a choice to stay unaware and see life differently:

L - Stands for LYING to myself, stay in denial, poor insight, lack of remorse, hostility are byproducts of lying.

I - Stands for IMPULSIVE and leads to trouble with self and others, leads to drug use, sex abuse and ends up with drug dependence, sex offender - a few of many problems.

F - Stands for FUN SEEKING in many deviant ways seeking short term pleasure and long term suffering.

E - Stands for END to peace in life, restless, anxiety, insomnia, depression, maladjustments are of many complications.

I will be a waste of my human resource, but will serve a great lesson for others.

Healing Tip 259

Truth without wisdom can be a weapon of mass destruction.

Someone asked a wise person how he had gained so much wisdom.

The wise person replied "From stupid people."

Healing Tip 260

Hate and love are the two most powerful forces humans attach to themselves and others.

These attachments can be dangerous and destructive if the reasons for these attachments are not healthy.

Self-love must be your top priority, and if any attachment can hurt you, then that is an unhealthy attachment that must go regardless of the reason or purpose.

Healing Tip 261

Humans were made with dust, by the Creator in His image with His own hand without any help. Then He blew his holy breath into human.

If humans follow the laws of Creator, then a human has the potential to be the Supreme Being.

If a human disobey the laws of Creator, then a human has the potential to be the devil being.

If I keep my body, mind, soul pure and clean and help humans who need my help unconditionally, Lord will reside in me. If I don't do that, the devil will reside in me.

Each human is an image of God with a Holy soul, if I see dirt in others, then I must check my eye for dirt and see other with Holy eyes to see their holiness.

Healing Tip 262

Mother wanted her daughter to learn French. The daughter was not interested. The mother hired a French speaking lady to tutor her hoping her daughter would be forced to learn French. After a couple of months she asked the tutor "Are you guys able to communicate?" The tutor replied "Very well, she taught me English."

Although this shows the ability to compromise, it also shows the lack of personal cooperation and growth between two beings.

Healing Tip 263

Special thanks to my wife of 33 years for being the inspiration to write this Healing Tip. Today is her birthday, so this is her birthday present.

MARRIAGE IS A HOLY WAR!

The one who captures and conquers the partner's heart is the winner!

A union of two hearts will transform into one holy heart.

This holy heart will pump love to nurture the body, mind and soul, and the winner will live a lovely life.

Healing Tip 264

On the stage of life, I am the judge, I am the victim and I am the defendant.

If my judgement is just, then my verdict will be fair.

If that judgement is executed justly for correction and transformation, I will serve my term well.

I will be a free person, not a judge, not a victim, not the defender.

Healing Tip 265

Less fortunate people are the greatest assets of any society.

If not taken care of properly, they will cause downfall to any society.

People give to the less fortunate and receive multiple folds in return.

People holding back not only do not receive more, but are also deplete and are not able to enjoy what they had.

If you hold back from the needy, then you will lose what you had and end up being needier than the needy you hold back from.

MORAL: Serve less fortunate with a smile, be happy and keep praying for more fortunate ones.

Healing Tip 266

You can conquer the world with a one-person army if that person is yourself.

Healing Tip 267

Time is the most precious commodity.

Proper use of time is the key to success.

Healthy sleep time will be followed by healthy awake time.

Unhealthy or lack of sleep will be followed by unhealthy awake time.

Healthy down time will follow the quality prime time.

Down time can be meditation, prayer, quiet time, soul searching or anything that helps one focus on self.

Religious books as well as non-religious books, scientific books advise us to do down time. For example, the bible tells to find a closet and be alone with GOD. One who is praying secretly will be rewarded openly.

Quran encourage to pray in the middle of the night for half the night or third of the night for self-purification.

Holy Hindu book encourages to spend time alone in the jungle or isolated place to acquire inner strength.

Medical literature found that meditation is beneficial.

Each human needs down time to have prime time.

Healing Tip 268

In the past whenever I was slapped, I slapped in return to get even.

This did not heal my pain, because I was not happy with the feeling I had afterwards.

When I could not slap the person in return, then I put him or her on my hit list, and waited to find the opportunity to get even.

I was not happy with the way I felt.

A wise person advised me to offer the other cheek.

I did that and not only ended up being slapped twice, but also I was laughed at for being so stupid.

Of course, I was not happy with the result, and did not like the way I felt.

I decided to do research and learn from my past events.

I was shocked that every slap was a reminder of my shortcomings.

I started on my short comings and sent everyone who ever slapped me a thank you note for teaching me the much needed lessons and reminders.

Now I receive a lot of hugs and thank you notes from almost everyone and live a great life.

Healing Tip 269

My conclusion of all these years of medical practice is that humans have four basic needs.

If those needs are fulfilled 100%, then a human has a potential to live quality life for 130 to 140 years and will die a pleasant natural death.

We are gifted with 168 hours per week.

The first basic need is a good night's sleep. On the average a human needs 50 hours per week of sleep. This leaves 118 hours.

The second basic need is healthy nutrition. Let's approximate it takes 14 hours per week preparing and eating. That leaves 104 hours.

The third basic need is exercise. It is recommended we exercise approximately 4 hours per week. This leaves 100 hours.

The fourth basic need is meditation, prayer or closet time which varies from person to person. Each one can gauge their need depending upon the quality of life. If you are perfectly happy and content, stick with the same time or you may need to increase or decrease depending on your needs.

The rest of the time is my NET time, quality time to work and feel like having a paid vacation or do rewarding volunteer community work, quality family time depends upon my need and live perfect happy fulfilled life.

Keep a timesheet and it will keep you for a long time.

Healing Tip 270

I asked one of my patients "How are you doing?"
He said - are you talking about I, My or Me?
I asked "Are they working like a team or is there a civil war?"
He said - how do you know there is a problem?
I said "I deal with this issue every day. I am still working on a peace treaty acceptable to I, My & Me. We know humans are the most sophisticated, complex and supreme creation."

We know the human has two systems running side by side not very compatible. Limbic system (equivalent to MY) primitive system, impulse control and cortex driven system or higher centers of reason control (equivalent to ME)

If a human receives impulse and is cleared by the cortex before acting on the impulse, that person will live a long healthy life.

Watch for the impulse that may determine your fate.

Healing Tip 271

I have only two eyes.
I must keep one eye looking inward at my shortcomings.
I must keep other eye on my surroundings.

Healing Tip 272

Religions of the east believe GOD resides in them, and they meditate to discover and stay connected with GOD to find peace and happiness.

Religions of the west believe GOD resides outside of them, and they pray to discover and stay connected with GOD to find peace and happiness.

BOTH ARE OK

The purpose of meditation or praying is self-purification.

Without self-purification neither GOD can be discovered, nor can we stay connected with GOD and there will be no peace or happiness.

Healing Tip 273

My past behavior is the best predictor of my future behavior.

Unless I choose to be brutally honest with myself, become self-aware and have the courage to confront myself and permanently transform myself to new and improved me.

Complete transformation is the most difficult and most rewarding task of human journey of life. Without complete self-transformation, the relapse rate to old behavior is very high.

Healing Tip 274

Humans have five senses and one brain.
I used to see problems and felt sad.
I used to hear problems and felt even sadder.
I used to talk about problems and turned myself into a problem.

Now when I see a problem I let my brain process it and I see opportunities.

When I hear something, my brain processes it and I learn wisdom and awareness.

Now my walk does most of the talking and I pay more attention to other's walks.

Now I am the solution instead of a problem.

Healing Tip 275

To have realistic expectations, I must constantly be updating my expectations like the stock market. My health stocks must do great, my quality and quantity of life depends on it.

Healing Tip 276

Each human is composed of 3 systems and a unique entity. All four of them make one entity that is a human.

Those three systems I may say I, My & Me.

'I' stands for the physical system.

'My' stands for my mental system.

'Me' stands for emotional system.

One unique entity within the human is MYSELF.

That entity carries a lot of traits including fear, anger, procrastination, jealousy, love, hate, caring and the list goes on and on.

First these three systems must work like a team before getting MYSELF involved.

There must be peace treaty between I, My, Me and MYSELF.

This may not be easy because MYSELF being my worst enemy and root cause of my issues.

I must love my enemy so my enemy will become my best friend.

MYSELF is like a wild horse, if not tamed it causes destruction to me and others, if tamed it will serve a great purpose.

Healing Tip 277

My late father wrote in his diary - I must close my eyes, ears and mouth to stay connected with GOD.

God will be my vision, hearing and speech.

In the past I could not quite understand what he meant.

Now I know what he meant. I read the story of human creation. GOD created Human and announced to everyone that Human is HIS masterpiece and will be HIS representative.

Angels questioned him stating that humans seem to be a defective product with anger, hate, impulse, procrastination and many other issues, and do not seem to have the ability to represent GOD who is kind, merciful, and giving.

GOD announced that HE knows the best and ordered everyone to bow down to human.

The Devil refused stating that he is much superior than human and it is unfair to be asked to bow down to the inferior product.

Now I know a human has two components: the humanly part made of dirt which has earthly desires and a love of worldly things, and the other part which is the spiritual and Godly component.

My father was a spiritual being, connected to GOD and represented GOD and was GOD's masterpiece.

Be a human and live pure and righteous life, never look down to anyone, treat everyone like a masterpiece and serve humanity, expecting the worse from them.

Healing Tip 278

I do work in the prison system and usually come across great and interesting encounters.

This encounter with one inmate really stood out to me.

I met with him on several occasions, and I was impressed with how well behaved he was.

I asked if he'd like to share his story with me.

"Only if I stay anonymous."

I promise that he would.

"Since childhood, I did the best for everyone that was easy for me, and I enjoyed doing it and felt wonderful.

Hard part was how people reacted to me, made me feel stupid and inferior.

One day I shared my feeling with my father.

My father said 'If you want to be pure & clean and stay clean, you will get harsh treatments, but those harsh treatment is directed toward the impurities within you not against you. If you are brutally honest you may discover that you needed even harsher treatments than you received for the wrong things you did. If you want to be pure and clean, and stay clean you need those treatments on a regular basis, unless you choose otherwise and be part of the crowd.'

I understood what my father said and I did what I loved to do and things got better.

When my father died, he inherited all his wealth to me.

Everyone was mad at me. I choose to wait until all of them cooled down and then shared my inheritance with everyone fairly.

To my surprise, I was arrested and charges against me were wrong and unjust.

I choose not to have a lawyer and not to defend myself. I did not want to embarrass my family.

I was sentenced to 60 years imprisonment.

I was kept in a single cell for 6 months and I discovered 7/17=24. (7 hours sleep, 17 hours prayer + 24 hours worship because I prayed in my dreams and I loved it.)

Due to my behavior and no write-ups, I ended up in populations. During my work-time, I stayed mindful and in a state of worship.

On off-days I continued 7/17=24."

I asked why he continued to do what he was doing.

He replied that there used to be 7/11 store but now there are AM/PM 24 hour stores. If you want to do good business, it cannot be 7/11 it has to be 24 hours.

"I am doing business with my LORD, I am well compensated.

I love it here, no distractions, not much opportunities to do wrong, provides me the atmosphere I need to self-purify myself. I am thankful."

"What happened to the inheritance?" - I asked.

"I gave all to my family, because I did not need it."

"Do they come to visit you?"

"No - That is OK and I do understand why.

I know Lord has plan for me so I do not have my plan."

Healing Tip 279

If anyone wants to work for the maker call 1-800-CREATOR. USE ONLY HEART PHONE. You will be hired on the spot with no requirements of sex, age, color, health conditions without any problems. No background checks. Great track record - no one ever laid off or fired. Ex-Employees were Abraham, Buddha, Moses, David, Saints and the list goes on. Request for reference check, Hired Immediately with no questions asked. People of any faith or no faith are welcome. Great benefits here and excellent after death package, palace in heaven all utilities paid and no copayment of any sort.

Job Description: Love, learn and serve the maker's creation. All shifts available, will accommodate your schedule, make your own schedule.

Call Now - You'll be happy you did.

Healing Tip 280

The only and only cause of disappointments and frustrations is having unreal expectations from myself and the creation.

Only if I get connected and stay connected with the maker will I then know exactly what to expect of me and rest of the creation.

Only then will I have real expectations, not only will I live disappointment free, and frustration free life. I may be pleasantly surprised that I and the rest of the creation are doing much better than I expected.

Healing Tip 281

Marriage is the union of two souls.

The souls are kept pure by both partners.

Humans do have likings, dislikings, desires, and those can be compatible or incompatible between the couples.

Whenever there is incompatibility, the soul driven person will search his or her soul and process in a healthy way, and the marriage will stay great.

They will be soul-mates forever.

Soul searching is the cure and not only it will resolve the incompatibility, but it will help the person to grow and preserve the marriage. Sooner or later the partner will fall in love and stay in love and both live happily thereafter.

If one of two is spiritual, that marriage will work. If both are spiritual, then it will be a perfect marriage. Marriage of two humans is mostly a struggle, or like business partners (Co-Dependent), and spouses cannot afford to live separate or mostly under construction with major delay or dead ends.

Unless they take a U-Turn and retrace their steps and discover where they took the wrong turn and get on the straight highway. To do all that one has to keep their navigator on (Their Soul)

IT IS NEVER TOO LATE TO TURN YOUR NAVIGATOR ON.

Healing Tip 282

Yesterday was perfect, today is more perfect, and tomorrow's perfection will be even better and so on.

Healing Tip 283

If you do not see the peaceful checks being deposited into your account, then you are not doing a good job of serving and/or forgiving others.

Healing Tip 284

There is a saying "All is well that ends well."
Every human being is an evolving story.
Every dead human being is a story, some with a happy ending, some with tragic ending, and some with unhappy ending.
If they lived their life well, they can avoid the unhappy ending.

Healing Tip 285

Time is like ice, if not used in a timely manner, it will melt away and be wasted.
Have a plan to live forever.
Have a dream and live it.
Live today like there is no tomorrow.

Healing Tip 286

I heard a statement "marriage is a major cause of divorce" - Let's analyze this.

Each human is born with a body and mind to stay together until death brings them apart.

We must stay married to our bodies, serving them as they serve us.

Healing Tip 287

I hear about New Year's Resolution.

Some claim it works, but the majority disagrees with this.

For example - Do I want to be happy and healthy? Am I willing to do what it takes to be happy and healthy? What I have to do anyway?

I am healthy and happy, and the plan is to continue being happier and healthier.

Others may have other commandments and that is fine with me.

Sometimes a dramatic change is not necessary, although change can be good.

Healing Tip 288

My walk should be louder and clearer than my talk.

I must pay attention, understand, and agree with my walk and keep my mouth closed.

If I cannot hear and understand my walk, no wonder no one will hear and understand my talk.

My mouth can be a weapon of mass destruction, and must be used wisely, like counting my steps of my walk, in the right direction and having peace with myself and others.

If my walk and talk are not the same, then I am a hypocrite.

Listen to other's talk and learn and keep improving your walk.

Healing Tip 289

The world follows the laws of nature

Nature fulfills everyone's needs and provides whatever is needed including human's needs.

Only humans have a choice to obey or disobey the laws of nature.

It may be knowingly or unknowingly, but the consequences and rewards are the same.

I have observed that only healthy, natural sex is for reproduction only. Any other form of sex is the root cause of many mental, physical, emotional and soul sicknesses.

Whenever I broke the law of nature I faced serious consequences, even my own body and mind rebelled against me.

Follow the laws of nature and live a perfect natural life.

Not knowing the law is not a valid excuse.

Healing Tip 290

One of my patients discovered the secret of happiness.

This patient I saw 3 yrs. ago was a very frustrated, single mother with 3 kids and 2 jobs that she was ready to quit.

She stated that she needed a new brain program.

I recommended to start with Brain program kindergarten 101

That is to eat healthy.

She stated I just do not have the time.

You flunked KG I replied.

Later she promised to eat healthy.

6 months later she came and told me that she graduated from kindergarten.

She was promoted to first grade which that is to exercise regularly.

KG is like putting the high octane gas, and first grade is burning that healthy gas.

The ride must be smooth, and if something is not right, then go back to KG.

She later came back and told me that she had passed first grade.

Healing Tip 291

Be in Heaven, whenever you want, you have the key FAITH

Safeguard the key, otherwise you may not be able to enter into paradise.

Healing Tip 292

If I have a healthy soul, God will reside in my heart.

Otherwise humans are just images and my heart will get attached to these images in a positive or a negative way, which is both equally toxic.

I will be the captive and live an unhappy and unfulfilled life.

Addendum: Feed your soul a lot more than you want to keep it healthy.

Healing Tip 293

I remember hearing "It is not the dog that bites the hand that feeds him."

I ask "Do you mean humans do that?"

Of course he replied.

"The dog that does not bite the feeding hand must be nurtured and tamed.

A wild dog of course will be different story.

Each human does have a dog within.

If the human is not being nurtured, of course the dog within will become a wild dog.

Like Hitler, Osama bin Laden, just a few of the many.

Humans who are nurtured well turn into a trained dog-people like Mother

Teresa, Edhi from Pakistan, Pope John Paul of course, just a few of many.

Lack of proper nurturing is the reason that many humans like me are not able to follow the very first commandment, thus not knowing that they have the potential to be the Supreme Being and live a perfect peaceful life."

Is the dog within you nurtured and tamed?

Healing Tip 294

The whole universe follows the laws of the CREATOR.

Even my body organ, the brain is programmed by the Creator.

If I choose not to follow the commandment of the Creator, I will then create trouble in the universe.

The whole creation including my body and mind will rebel, and will be very difficult because I will go against natural flow.

Obey the creator or else!

Healing Tip 295

God made the flesh from dirt.

God built the human heart as God's residence.

That made the human body very special.

Then God put the soul (part of God) in the human.

The soul made the human the best creation.

Each human is a soul mate to each other, because we all share the same soul.

Each human has a soul and a potential saint.

We are not body or flesh mates.

God resides in each human heart. The heart is God's residence, so please keep it pure and clean.

It is not a guest room.

Healing Tip 296

God is always listening, but he is not judging.

Healing Tip 297

In the old days, rivers were used to transport wood logs.
Two wood logs were floating in the river.
Log#1 noticed that when it jumps all of the river jumps and when it floats quietly, then the whole river floats quietly.
Log#1 told log#2: "I am the master of the river, and I control it."
Log #2 replied God is the master of everything including the river.
Log #1 said that log #2 is just stupid and cannot comprehend log #1's greatness and does not understand the wisdom.
The people caught those logs. Log#2 lived naturally and was in good shape, so they made furniture from log#2.
Log#1 had lived unhealthy life and was not in good shape, so he ended up being used as fuel.

Healing Tip 298

When I reflect on my life, everything I did knowingly and unknowingly served me well.
*Sometimes I thought I was doing the right thing and did it, then realizing later that it was not the right thing.
*Sometimes I knew something was wrong, but I did it anyway.
*Sometimes I knew what the right thing was, but was unable to do it.
*Sometimes I did not even know what the right thing was.
*Knowing the real right always has been harder than doing it.

Healing Tip 299

*If I pay attention within and out of me I see a lot of systems are working, some simple, some complicated and some I cannot comprehend.

* Some people and systems work very well together and some do not.

* I too work well with certain things and people and with some things and people I cannot work well.

*If my connection with God is perfect, then I will see the hidden wisdom in everything and everyone, and live a purposeful life.

This is the End of the Trail of Healing 303 - Lesson #3

If you have successfully climbed the trail, precede to Lesson #4 - Healing Tip #300

GOOD LUCK

Healing Tip 300

Your mind, heart, and body are houses of God, and you are a janitor. If you do not keep everything pure and clean, then you may suffer serious consequences like depression, anxiety, heart attack, stroke, and may damage the vital organs.

This may be permanent, and it is never too late to ask for help.

Healing Tip 301

One time a man went to Moses and asked Moses to tell God to make him the richest man and to make him live forever.

Moses conveyed his message to God.

God said I will grant him his one wish, but not both.

The man chose to become the richest man.

After many years Moses saw him still alive, and he asked him what is keeping him from dying.

The man replied "I made sure that I gave so much charity to those lesser than me, that I truly am the richest man."

The more you give, the longer you live.

Healing Tip 302

* LAW OF NATURE*

Do what you want to do.

See what want to see.

Say what you want to say.

Feel what you want to feel.

Treat others the way you want to be treated.

Treat your mind, body, and soul the way you want to treat them.

Everything and everyone will do the same to you.

Healing Tip 303

In 1992 I was doing my residency and used to do history and physical exams for psych patients.

I remember seeing one patient whose state of mind was very bad.

He described that bad thoughts made him so uncomfortable, that he wished he was dead, and was then placed on drugs. This made him feel like a zombie.

"There is no happy medium. I tried to kill myself, talk about being a failure, I did not even succeed in killing myself. God is punishing me for the bad things I did in my past. I pray to God please take my life and that will be the end of my miseries."

I suggested to chant: "God forgive me, God help me all the time."

I saw him 18 years later in 2010 in a clinic, I could not recognize him, but he recognized me.

He said:

"You saved my life, and now I am working and I chant all the time "God forgive me, God help me," I do it all the time and life is good.

God is the universal healer.

Healing Tip 304

There are some people in your life that you cannot choose.

These include your parents and siblings.

But, with most other relationships, you have an ability to choose you want and do not want in your life.

Make sure they want to be in your life as well.

Healing Tip 305

If I work for GOD, I must do the best for the creation, so I can get my pay check, and expect the worst from the creation, in return.

I must have a strong defense system in place that creation cannot hurt me, so I continue to serve creation and continue to receive my rewards from the Creator.

God has an opening for you, and it is up to you to sign a contract and fulfill your obligations, so you can receive your pay with benefits in this life and life after death.

Healing Tip 306

Do I have a purpose? Do I have a mission statement for myself?

After thinking long and hard I realized that my purpose is to submit to

GOD and my mission statement is to Serve God's creation.

Healing Tip 307

Giving birth seems fearful and painful, but so is old age and death.

The more pain you are willing to bear, then the more joy I will have.

You should be willing to embrace death, knowing that death is the thing giving you your purpose to live.

Not willing to bear pain will lead to a purposeless life and fearful death.

Healing Tip 308

I liked what I read somewhere that the speed of light travels faster than the speed of sound.

This is why some people appear bright until they open their mouth.

In reality some people do not have to open their mouth, because their actions are much louder than words.

That talk travels much faster than light and can be very effective or destructive depending on who receives it.

Everything lies in the eyes, ears, and other senses of beholder, depending on how it is processed.

Healing Tip 309

Some believe that birth, marriage and death are preordained.

My personal believe is that everything is preordained and designed to teach me much needed lessons so I can grow and be on a honeymoon with myself until death makes me part.

Nobody has a choice about their birth.

Marriage and death seem like mysteries since nobody can predict what happens after them. So it is understandable why people are so afraid of them.

But death cannot be avoided.

Healing Tip 310

No pain, no gain. Pain without gain is unhealthy and unnecessary.

No suffering, no joy. Suffering without joy is unhealthy and unnecessary.

No trash, no cash. Trash without cash is unhealthy and unnecessary.

No work, no paycheck. Work without paycheck is unhealthy and unnecessary.

No growth no wisdom. Growing without wisdom is unhealthy and is a waste.

There is always two sides of a coin.

A coin with one side is worthless.

Healing Tip 311

One person was frustrated that he was always back stabbed by his friends and family.

He wished that he could see the true intentions of humans.

Then, a genie showed up and gave him a magic cap, which when put on his head would show the true identity of everyone.

The man was delighted and so he put the cap on and was surprised to see his best friends and family resembled pigs.

All the people he used to look up to looked like devils.

All the people he used to look down on looked like angels.

And when he looked in the mirror, he saw a pig wearing a cap.

If you do not like the company you keep, change yourself and the company will change.

Look into the mirror before you look at others.

Healing Tip 312

All my life what I wanted, what I needed, and what I received never matched.

Now looking back I realized, that I wanted, received, and needed was a perfect triangle and the world was always perfect.

Now I can say with certainty, my past was perfect, my present is perfect, and I know my future will be perfect.

*God is in charge, and God will know what was done was perfect, what is happening is perfect,

and that your future will be perfect, because He who is in charge is perfect.

Healing Tip 313

The devil judged the human flesh and was rejected. He was unable to see GOD in humans.

When any one says or does anything to me, if I ask GOD to tell me the hidden message, then I will hear the word of GOD.

If I believe my ears and process through my brain, the message I hear and understand is my flesh understanding, which may not be the true message.

Healing Tip 314

Humans are created in the best form, and are sent here to Earth which is a boot camp.

The goal is if you walk on a pure path, and your boots must stay clean and pure, then you will reach your destination: Heaven.

If you notice that your path is not pure, or your boots are not pure or clean,

You can get new pair of boots, and start your path on pure straight path.

Boot supply is limitless and it is never too late, to get on straight path.

Healing Tip 315

It is easy not to care about yourself and others.

It is lot harder to keep your opinions to yourself.

Watch that tongue, It very well can be a weapon of mass destruction.

Healing Tip 316

Training on Earth helps humans to learn to deal with themselves, others, situations, and be able to relate to everyone and everything.

The goal of this training is to know the good and bad.

Humans have a choice to graduate any time from this training once humans feel they have acquired their goal.

After graduation everyone has a choice to be good or bad.

Basic training does not have strict time limits, any time a human can choose to graduate and make the choice.

Healing Tip 317

Today is a new year, and it is a cross road. How did I get where I am?

If I review my life, I find lots of people did nice things to me as well as bad things.

I did good things and bad things to others: I did good things to myself and did self-abuse.

What have I learned from my past??

How is my past helping me to have a better present?

People did nice things to me; I must continue to do nice things to others and be thankful.

People did bad things to me; I must become a strong advocate against those things and be thankful that I learned my lessons.

I must thank GOD and be thankful.

I need to confess the bed things I have done, and beg forgiveness from the Creator.

I must do community service to neutralize the toxic effects on self-abuse.

I am guilty and ready to serve my life term to serve humanity.

Healing Tip 318

This world is a prison and we are all on death row.

The golden rule: "Judge not and I will not be judged."

If I look down on anyone, I will follow devil's footsteps and will be rejected.

If I learn from every one, everyone will be my teacher and in return I must serve my teacher and earn by rewards and I will serve my time well.

I will live a peaceful life and death will be the door to freedom to go back home and there will be a life after life in heaven.

Judge others and this prison will turn into a torture chamber and you will be afraid of dying and die a painful death and may end up in a hole.

Healing Tip 319

What is my disease, or block, that is stopping me to live the kind of life I want to live???

I have to go back to the root cause.

If I believe that God has created me, other people may have other ideas,

But I have more reason to believe that God has created me and I have to serve a purpose. My question is what is my purpose?

Maybe searching for my purpose is my purpose.

I discovered what I was looking for, it is simple, I am here not to judge but learn, serve, and love.

I learned that I must love my enemy so I will have no enemy.

I discovered I am my worst enemy, so I should start loving myself and realize I have no enemy.

I discovered that I must love those whom I was supposed to hate the most.

My best teachers are those I never met, but I know they are my worst enemies: the devil and my narcissism.

The devil is very talented but his untamed narcissism made him a loser.

I do not want to be a loser like him. I leaned the following lessons.

*I must never have an untamed narcissism.

*I must never judge and never look down to anyone, but judge myself.

*The devil is determined and working very hard, his mission statement is divide and rule.

*My mission statement is love, learn, and serve, and I must work as hard if not more than him.

I cannot see him but I hope he may read my tip and realize and repent and become a winner.

So I can return his favor and learn from each other. We are all teachers and student at the same time.

Healing Tip 320

Someone said if I am not with the one I love, and love the one I am with, then I will not be happy.

But I do not think any human has the ability to do so.

Let's analyze this. If I am not with the one I love, that is a message from my Maker that I am following manufacturer instructions properly.

I am a gift from my Creator to me, and if I do love the gift and be thankful to the Creator, I will love myself the most, and love & serve the creation.

Healing Tip 321

Special thanks to a very special correctional officer to teach me the needed lesson.

On Tuesdays I go to a local prison to conduct religious services. I am supposed to be there at 5.30 to distribute sheets to a different block and start services at 6PM.

I am a procrastinator, so I reached there at 6 pm and saw this wonderful C.O. and I gave him my healing tip #321. I knew he liked that kind of material.

At 6.30 when I reached the chapel, the same C.O terminated the services because the inmates could not be alone in the chapel. I saw God's image in uniform telling me procrastination is a sin and I have committed it.

That was the answer from God for healing tip #321 who gave me the ability to see God's image in humans.

On that day I was a weapon of mass destruction, causing trouble in prison, and depriving about 35 inmates from the services instead of being a tool of construction.

I owe apology to the C.O and inmates for my major sin and I am thankful that I learned my lesson.

Procrastination is a sin and the person who procrastinates is a hypocrite and the root cause of procrastination is narcissism, a very complex system in humans and if properly tamed and harnessed, can turn humans into weapons of mass destruction. If properly disciplined, tamed, and harnessed, that person can confront any evil force and become a tool of mass building and reconstruction.

To tame narcissism I need to go to prison to learn one of my lessons.

Healing Tip 322

Special thanks to Jasmine Sample whose heart is crystal clear and saw God's image and inspired me to write this healing tip.

I was made in God's image.

If I do not see God's image in the mirror, then I must be able to see the thick covering of my sins covering that image.

If I do not see that, then I am nothing but dirt.

If I do not see God's image in others, my eyes are covered by my sins.

Healing Tip 323

Special thanks to a kind human being, Shannon Quigley, for being my inspiration to write this tip.

Life teaches each human being the much needed lessons which humans can learn or disregard. What they do with these lessons shape their thinking process and how they interact with themselves and others.

Religious people are taught to be religious and how they should view others. This is okay as long as they are open towards other beliefs as well.

Of course not every religious person practices that way

Psychiatrists are taught in med school and during their training how to diagnose mental issues, and to treat them only if they are being paid. Not every psychiatrist practices that way.

There are those exceptional people who can see under the surface and learn from everyone without judging and serve everyone with best of their abilities.

Those people take the path of lifelong, effortful learning, love, learn from everyone, and only teach if they are asked to do so whether the other has the ability to pay or not.

Those people transform into righteous people, with a sense of purpose, and acquire a peace and happiness which can only be felt, and is very difficult to explain to others.

Healing Tip 324

BEAUTY IN THE BEAST

A long time ago there was a beautiful princess, not only was she beautiful, she was more beautiful inside, and she was humble and righteous.

She was an angel in a true sense.

There was a beast in the kingdom who used to rape and kill women.

One day, that beast got into the palace and grabbed the princess.

The princess looked him in his eye, and the beast was shocked.

No one ever dared to look into his eyes like she did with not fear, but love for him.

He started trembling and fell to the ground.

The princess said: "I see a beautiful soul in you, let's pray together.

I will pray for your forgiveness and you repent for your sins."

He agreed, both prayed, tears flowed like streams, and their clothes got soaked.

Suddenly, the princess stood up and kissed the beast on his forehead and he turned into a prince.

Both thanked GOD, and the princess asked him to marry her.

He replied, "I have to serve time for my crimes. My repentance will not heel the pain I caused, or repair the damage I have caused.

Maybe I can do community service the rest of my life."

"Why not join the prison ministry?" The princess suggested.

He agreed and filled out the application to work as volunteer in the prison ministry to help inmates.

He applied at a local prison to be a volunteer priest. They did a back ground check and discovered that he was a murderer, rapist, and got arrested for rape and killing charges. He was sentenced to a life term.

However, he was allowed to conduct services for male inmates only but with, of course, not pay or benefits.

He accepted the position and soon he prayed for everyone, and inmates who repented started turning into righteous people.

Those righteous inmates ministered to others and very soon the prison turned into a paradise.

When the princess visited the prince (ex beast) he requested her to minister female inmates.

The princess agreed and the female inmates started turning into righteous women.

When these inmates got released, they would minister others.

He learned a lot. He discovered innocent inmates serving time and he reminded them that they are following the footsteps of Joseph, who was also innocent and that comforted them.

Very soon there were no crimes in the kingdom, no prisons except one which housed the princes.

He continued doing 7/17 7 hours sleep, and 17 hours of worship.

Everyone in the kingdom lived happily ever after.

The princess rented a room across that prison, and faithfully visited the ex-beast every Thursday 9 to12, the only time she could.

Both lived happily even though their bodies were apart, but they were true soul mates and both died the same day and were buried next to each other and lived in heaven together forever.

Healing Tip 325

Last Sunday in the Harrisburg newspaper there was story about a 25 year old marine who served two combat tours, one in Iraq and one in Afghanistan.

When he returned home, he faced more viscous internal enemies of fear, guilt and hopelessness.

He had no training to fight those enemies, and all the weapons he tried, such as counseling, and medication, and sheer will failed him. He successfully killed himself and lost the life game. Sadly, he died before his time.

He was diagnosed with PTSD (the war within) caused by bad experiences.

There was another event in Texas where a person killed and wounded multiple people, and then killed himself.

It seemed nothing worked for him either.

Each human goes though PTSD in one form or other; that is the part of growing up.

What equipment do you need to combat your inner war?

Healing Tip 326

Once upon a time long ago there was a very powerful man with anger issues.

He used to kill anyone who ever mad him angry.

After killing 99 people, he felt guilty and started repentance for his sins.

He went to local religious leaders and asked them if God would forgive him.

"No way!" the religious leader replied. That made the man angry and he killed the religious leader.

He felt guilty and started repenting again for his sins.

Someone suggested that he can go to Moses and ask him to ask God about being forgiven.

He made a long journey and met Moses and requested him to ask God if he was forgiven.

Acknowledging all sins and wrong doings and true repentance will turn all into good deeds into a blank slate to start new sin free life and live happily thereafter.

Healing Tip 327

If you read lips, listen to body language, and let people read your lips, then you will learn from everyone and live a peaceful and happy life.

Healing Tip 328

Humans are flesh made of dirt, and we will go back to dirt after serving our purpose.

Our purpose is to house and carry the soul like a horse, and the soul will be the rider.

If the flesh is not tamed properly then it will take the soul where it is not meant to be with serious consequences.

Flesh driven person can be comfortable and unhappy.

Soul driven people are always happy.

My real war is with my flesh, taming it and keep it under control.

Healing Tip 329

Time is money. This is not true. Time is priceless. It is the most precious limited asset, and I do not even know how much time I have.

When my time will be up, I cannot buy a moment with all the money on earth.

Capture the present moment and use it the way it should be, and it will turn into building blocks for the future and will continue the same until the end of this life.

All the wasted moments of my life are turned into toxic waste, and the earlier I recognize that the better.

All the toxic waste needs to detoxicate and recycle.

Otherwise, the toxicity of my past will never allow me to capture the present moment perfectly.

Wasted moments turn into wasted life which turns into toxic waste. This waste becomes too toxic to be buried or throw into the ocean because it will make the whole earth or ocean toxic.

The only safe way to rid of this waste is to put it into the furnace.

Healing Tip 330

If I see good in others, then I should pray to God to increase their goodness, and try to acquire that goodness.

If I see evil in others, I must pray to God to remove that evil, and watch for the evil within myself so that when it is found I can either get rid of it, or tame it and place in a cage.

The day I will discover the evil within, tame it and cage it can only happen with the help of God.

If I am unable to discover evil within, my life sooner or later will turn into a living hell.

Healing Tip 331

All my life I was looking for saints.

Every time I thought I found a saint it lead to disappointment.

I thought that there were no more saints in our time.

To my surprise, I started discovering saints in prisons, hospitals, and among my patients.

That was God's way of telling me that I was not looking in the right places.

One of my 32 year old girls, who hurt her back, came to see me.

All my life I prayed for everyone because I always felt I was the most fortunate person.

When I got hurt I did become selfish and prayed for myself, and felt guilty.

I realized this is a gift from God, and now I have more time to pray for others and I should be thanking God for this gift.

I saw a saint, a pure heart with healthy feeling who was sending healthy thoughts to the brain so its brain will not entertain any unhealthy thoughts, and a body who only does the healthy righteous acts.

Healing Tip 332

Lori Duran's rules of life

1) A live thinking person keeps changing and transforming with time.

Rocks and stones do not change.

2) Difficulties are gifts of God so humans can understand the purpose of life.

3) The true feeling of a soul is peace, and happiness is a temporary feeling.

4) Have no regrets in life, just lessons learned.

5) The secret of change is to focus and use all your energy, not fighting the old, but building the new.

Healing Tip 333

Kia Washington's rules of life:
1) The bible is the only refuge for a Christian.
2) Understand the message and practice it faithfully and follow Jesus' footsteps.
3) They quest of knowledge in any form is a form of worship.
5) The first step towards righteousness is to earn an honest living, and light of faith stands on the foundation of honest living.
6) Work and serving creation is a form of worship.

Healing Tip 334

Joe's MOTTO of life
ENDURE
Mon: E Embrace each day as a gift.
Tues: N Never surrender except to God.
Wed: D Do not let your situation get the best of you
Thurs: U Understand God's purpose and presence.
Fri: R Remain as positive as possible.
Sat: E Exercise mind, body, and spirit.
My addition SUNDAY: reflect, repent and rest and get ready for Monday to repeat the cycle, again and again.

Healing Tip 335

Saints were around me, but I did not have the vision to see them.

Some saints remind us of our shortcomings, and some give us the solution to our problems.

A couple years ago Saint Denita Henson asked me if I have any skeletons in my closet.

I reflected for a moment and replied "There is a graveyard in my closet."

Later I felt lots of regrets and that feeling continued for a couple years, till Saint Lori Duran wrote in her rules of her life and one of them was (No regrets just lesson learned).

It answered my question: "If I have a graveyard then I should be happy that I have learned more lessons than anybody else."

Thank you Saint Denita Henson for showing me my trash.

Thank you Saint Lori Duran for helping me turn to my trash into cash.

When is the last time you have checked your closet???

Healing Tip 336

When I start reflecting on my life, I realize I did some good things for myself and others.

Others did good things and bad things to me.

I realized I am made of dirt, and I am inclined to do dirty things, and by following that impulse I did bad things.

Others have done bad things due to those dirty impulses too.

But I did good things, and others did good things to me.

So there are angelic parts in me and others.

Everyone has the choice to show off their angelic parts or their dirty parts. The choice is up to you.

Healing Tip 337

Rev Jacqueline Keys Rules

1) Difficulties are a gift from God, so humans can understand the purpose of life.

2) The grace and mercy of God also is an example & testimony of that mercy and grace.

5) The secret of life is where ever I am, whatever condition I am in, so be happy.

Healing Tip 338

This tip is for a very special person Denita Henson on her marriage.

Denita & Gerald were granted separate bodies, but only one soul.

Human bodies need upkeep and maintenance to function well.

The soul also needs exercise diet and maintenance.

If I do exercise, eat healthy, diet, and sleep well, then I will live full life and my vessel will carry the cargo to my destination.

To have a healthy soul mate, not only does my body need to be healthy, but part of my soul must be healthy to help my soul mate when the need arises.

Two bodies and one soul need a lot more upkeep than one body and one soul, but the reward is many folds more.

The one who will carry the more weight will turn into best half.

Denita do not settle for just being a better half, turn into the best half.

Denita you can do it!

Healing Tip 339

This body and whatever it contains is a loaner to me by GOD.

God may repossess this loaner anytime, and it will be inspected thoroughly to check for any abuse.

If any abuse noted, then I have to serve my time in hell.

Self-assessment is necessary for daily bases.

Self abuse, abusing others, and allowing others to abuse you are major crimes and will be punished severely.

Anxiety, depression, blood pressure, high cholesterol, diabetes, and not feeling good could very well be early signs of self-abuse -abusing others or allowing others to abuse me.

If you have any of the above mentioned you may the victim of self-abuse or guilty of self-abuse,

Call your doctor now to get an evaluation.

If you feel you are not guilty or sick please call your lawyer to defend you in the court of GOD.

Healing Tip 340

As I learn more, I realize how much more there is to learn.

The more I learned, the more unanswered questions I had.

But also, the previous answers were wrong answers and conclusions.

To my surprise, I was happy that the answers I thought were the correct ones turned out to be the wrong ones because they were great learning experiences.

One day I was reflecting, and I realized that the reason I was getting all the wrong answers was because I was using wrong passwords.

Accidently I used password God1 and all my questions were answered.

Now I realize everything did happen to me to others were perfect and I must love and learn everything and everyone.

Healing Tip 341

God is the master planner, and everything and everyone is following God's plan.

God is perfect and His plan is perfect.

One question, what is God's plan for me?

When I reflect on my past, present, I realize I always had a beast and angel within me.

I committed crimes and sins under the influence of a beast, and did good deeds under the influence of an angel within.

Angel food is praying, helping, and loving everyone and everything.

Talk less, eat less, sleep less, pray a lot, and exercise a lot, as this will help with self-purification.

Excess eating, sleeping, premarital sex, are extramarital sex is food for the beast within.

God gave me a choice, choose A and be an angel I win, choose B and be a beast I lose.

Healing Tip 342

A Rabbi and a Priest got into an argument.

The Rabbi claimed only Jews will go to heaven, and the Priest insisted only Christians will go to heaven.

The Priest said "I bet you have eaten ham."

The Rabbi said "Yes I have eaten ham, but I bet you had committed adultery."

The Priest said "Yes not only did I eat the ham, but I am forgiven and you are not."

The Rabbi said "I am the chosen one and I am sure that I am forgiven."

The Priest disagreed, stating that the Rabbi will be forgiven only if he accepts Jesus as his savior.

They went to the judge. The judge said "Eating ham is not against the law and as long as adultery is not reported and proved, it is not against the law."

Priest disagreed and they went to a Buddhist monk for his opinion.

The Monk said both the Rabbi and the Priest should confess to their congregation

They both got mad and filed a case against the Monk saying that he had been abusing his wife, and he ended up in prison.

That Rabbi and Priest agreed that eating ham is not against the law and as long as adultery is not reported it is not a crime.

They also agreed as long they are law abiding citizens, that they both will go to heaven.

The Priest was not sure if the Rabbi would like being in heaven, because it would be filled with Christians.

The Rabbi was not sure if the Priest would like being in heaven, because it would be filled with Jews.

They also agreed spouse abuse is a crime and should be severely punished.

The Buddhist monk reflected in prison and realized he was a sinner, and a sinner cannot pass judgment so he appealed for forgiveness, stating he would never abuse his wife, and apologized to the Rabbi and Priest.

The Rabbi and Priest realized they were wrong and they confessed in front of their congregation.

Both congregations admired their courage.

The Buddhist monk was free, and so were the Rabbi and Priest.

All agreed that only God knows who will go to heaven, and everyone lived happily ever after.

Among humans who are believers, Jews, Christians, Sabians (any faith or no faith) if anyone believes in God, and on the day of judgment and performs good deeds-they will be rewarded and will have no fear & no grief.

Do not be like those who preach on pulpit to show off, like hypocrites, they do have their reward.

Healing Tip 343

SAINT IN THE PRISON

I work in the prison and discovered a saint.

After a couple of days he asked if he could share his story, but wanted to make sure he could remain anonymous.

I assured him that no one will know, and that I will use the name of Joe Smith.

It was a very long story, very inspirational with great lessons. This is from Joe's perspective.

I grew up in a small town with big family.

We were very poor, my father was a farmer who worked long hours and was too busy to help others and was not paying attention to his own family.

My mother and father both were very nice and hard working.

My elder brother started having sex with me when I was 9 years old and always paid me, so I never thought it was wrong.

His two friends started having sex with me and paid me and it continued.

I never thought that something was wrong.

When I turned 15, I started doing that to my younger brother and other kids, and never thought that anything was wrong.

I was bright, and I got into college, majored in business, and started my own business and had kids.

I continued my promiscuous sexual behavior even after marriage and never thought there was anything wrong.

Life was good. Due to my accountant carelessness I ended up in prison and met other inmates, and realized that I was not very different from them.

I am fortunate that my father and mother provided me with a stable home, and were great role models.

If I would have born in a different home, then I would have been their shoes and doing worse than them.

For the first time in my life, I had time to reflect; I remembered my father always tried to teach the Bible,

I could not really understand.

I started reading the Bible; it seemed I was reading the first time.

I loved the verse Matthew Worship in secret, God will reward openly.

I truly practiced that, and I started remembering sins I had forgotten and I repented and repented.

Sometimes this struggle with myself became so painful, that I wished to be stoned to death for my sins.

Finally, I started feeling a certain kind of peace I never experienced before. I achieved peace with God.

That is the best thing that happened to me.

I am repenting all the time and I know how I will spend rest of my life.

My conclusion: DO NO WRONG.

Especially sex which is meant for reproduction only between husband and wife.

Any other kind of sex is toxic to the soul, even lustful looking on others, even pictures are equally toxic to soul.

I realize it and never ever do it again and become the strong advocate against the one I had committed.

Yesterday I was a devil, and I did not know it,

Today I am a devil and know it. I rather stay in cage (Jail) until

I can cage the devil within so I can live my life.

This is a most difficult task I ever had to cage the devil within.

Hopefully tomorrow I will be a Saint.

NOTE: Each Saint has a past.

Every sinner has a future.

The devil has a great past, but no future unless he repents and is forgiven.

Healing Tip 344

Long time ago there was a small town where it had not rained for several years

People were praying for rain because it was so dry.

One man saw a poor man praying stating "O God forgive our sins and give us rain" and continued repeating "give us rain, and give it to us now and keep raining."

To that man's surprise, it started raining and he was impressed.

The poor man stood up and walked away, and the other man followed him to his home.

He gathered information about that man.

He was told he keeps to himself, and that no one really knows much about him.

He decided to investigate and discovered that he works one day a week with a temp agency.

That poor man showed up to work and his employer made a great offer: a lot of money and offered him to stay in his house.

The poor man questioned why he was so generous with him.

"Is this important?" he asked.

"Yes, it is the most important thing I'd like to know."

That man told him the story of rain.

That man said "I cannot accept that offer," and said "you must be a righteous man otherwise God does not reveal this kind of secret to anyone.

I had secret relations with God and life was good, now you know this and then others will know it. I do not want it and do not need it."

He started praying right there, "God take me now, take me now" and before the other man approached that poor man, he was dead.

People made a huge grave for him and people still go there to pray.

Righteous people are special people of God and God does reveal that secret to very special people.

Righteous people are all over sometimes they do not know themselves.

You might find one in the mirror.

Healing Tip 345

Moses was with his people in the wilderness.

People were driving him crazy. They wanted water nearby instead of fetching water from distant places.

Moses prayed to God and God said "Hit the rock with the stick. Hit it once and wait."

People were yelling and screaming, Moses was getting upset. He felt like hitting people on their heads, but he hit the rock instead and waited. The crowd started making fun of Mosses, one said: "you didn't hit it correctly hit it again!"

Moses was waiting, but the crowd was getting angrier and angrier, so he hit the rock for 2nd time and there was a great flood.

God said to Moses: "I told you to hit once and wait."

Moses said "People are driving me crazy. I hit the rock the 2nd time and broke the commandment and I am sorry."

God said "You were impatient, you were angry toward my creation so you will never see the Promised Land."

Have patience.

Healing Tip 346

God is #1
God created me and blessed me and I am #2
God also created everyone and everything and that is #3
Now I know my 1, 2, 3...
Come on every one please sing with me.
I am praying for everyone, everyone please pray for me.

Everyone who has helped me grow, please write what you think of my tips and me.

Now I know my 1, 2, and 3.

Healing Tip 347

I work in local prison as Muslim Chaplin and other prisons as a Physician and I come across great people with great stories and great insight.

One Muslim inmate who is waiting for trial for murder charges, shared his story, and he showed me the picture in the newspaper and stated that is not him, that the devil had taken him over at that time.

"I am asked to confess and get life term or go on trial and may end up on death row.

I reflected and realized that I need to consult God.

I know this is a reminder from God like many in the past, which I ignored over and over.

God has made it clear that this is the best place for me to get to know God, myself and my purpose.

God is aware of the crimes I had forgotten and many I did not even knew were crimes.

I am asking God to decide what I must do.

I am feeling peace I never experienced before, and I'd like to share this peace with everyone.

I also feel that God has forgiven my sins, and I must be the strongest advocate against the very sins I had committed, try to neutralize the toxicity I caused."

God is waiting for us to reflect and repent for our sins, and God will turn our sins into good deeds and reward and bless us. WHAT WE ARE WAITING FOR?

Healing Tip 348

Religion is like a knife.

In the hand of some religious people, it is a tool to defend their religion, they would even justify killing someone.

In the hand of righteous or spiritual people, it is a tool to slaughter their own desires like Abraham did.

Some religious people see wrong in the world.

Righteous or spiritual people see wrong within.

Some religious people's talk is better than their walk.

True righteous people's walk does the talking for them.

Righteous /spiritual people forgive and always win in the long run.

Religious people wish everyone would follow in their religion.

Righteous spiritual people wish everyone would have better peace than he or she has.

Religious people are at war with the outside devil of them.

Spiritual/Righteous people are with the devil within 24/7.

Healing Tip 349

God created all the universe and angels and everything and everyone just obeys God's will.

God created the devil and gave him a strong will.

The devil has a will and he chose to obey God's will and he became the chief of angels and felt good and was hoping that God would appoint him as his representative.

Then God created humans and gave them a weak will and declared that humans would be his representative.

The angels questioned God, and wondered why he had created humans and had given them power because they believed humans would only create trouble and bloodshed.

God replied: "I know what you do not," and ordered everyone to bow down to Adam and Eve.

All did except the devil. The devil said "I am better than humans and it is unfair on your part to ask me to bow down to inferior creations, if you give me a chance I will prove it to you."

So the human saga began.

Humans are no match to the devil, but any human will submit his or her will to God, and will become the supreme creation.

That is God's purpose for us, and we should have empathy for everyone including the devil and learn, love, and serve everyone.

Healing Tip 350

Special thanks to Saint Qudsia Siddiqui for being the inspiration for me to write this healing tip.

God is #1, I must know God and obey God.

God will make me aware that I am God's gift and that everything I have is a gift.

God will teach me how to use these gifts properly.

That awareness will teach me my abilities and make me realize the sickness of heart, body, mind, and my soul.

God will help me to live a healthy life.

When I will live a healthy life, then God will grant me the ability to serve creation, and I will receive my compensation from the Creator.

Sickness is a gift from God to make me humble, so I can have empathy with others, and learn to live a healthy life and serve my purpose. (Eternal job security)

Healing Tip 351

Special thanks to Pat Williams for being the inspiration for me to write this healing tip.

There are several theories about the beginning of human race.

1) Theory of evolution: Human was primitive and evolved to be a human.

2) Reincarnation theory: depending what humans do during their life time, come back as other forms of creation; some time rat or as animal or as a human.

3) Bible & Quran theory: God created human and declared that humans are the best creation.

Everyone questioned that statement.

God's claims that humans are better than the devil.

The devil claims that he is better than humans.

I have empathy for the devil. He made a grave mistake due to not having empathy for humans and could have learned from humans and could have live happy peaceful life in the company of AL MIGHTY.

The devil looked at the outside of humans which is just a vessel, unable to see the cargo (soul) that make humans the best creation.

Having empathy for everyone is learning and growing process that makes humans the best creation.

In my life whenever I felt sympathy for others, I was always paid with grief or frustration because

I judged that others needed my help, instead of having empathy and learn and teach each other and being paid with peace and joy.

Empathy I win.

Sympathy I lose.

We are human being because we have a body, and we also are spiritual being because we have a soul.

Healing Tip 352

A long time ago there was a merciful kind LORD, almighty & all powerful.

The almighty LORD created the heavens, angels, animals, plants, and the universe.

Everyone was living in peace and harmony- no one had free will.

Every one called the LORD big "L"

Lord created Lucifer with fire, and granted him free will.

Lucifer choose to obey the LORD and excelled and was appointed as chief of angels.

Everyone started calling him little "l" and he ended up having a big head.

There was peace and harmony everywhere.

Lord created Adam with dirt, and instilled a soul in him, and declared that Humans would be

LORD's representative.

Lucifer refused to bow down to Adam.

Lucifer said LORD I am made of fire, Adam is made of dirt, I am better and I will prove to you if you give me the opportunity.

That was the first sin.

He did not obey the lord's command, and therefore never found out the supreme secret.

Humans are the best creations because they have a soul.

Healing Tip 353

There are 10 famous commandments.

Some believe in those, some do not, some agree with some commandments and some disagree with other commandments.

When I reflected on myself, I realize I had my own commandments.

The first commandment is to worship one God.

I love money, family, country, desires, business job, friends the list goes on.

Do I love these things or worship these things?

I agree with the first commandment, but I worshiped many gods.

The first commandment is like first grade to me.

If I am not worshiping one GOD then I am failing my first grade.

I decided to worship GOD, be mindful of GOD.

I really had to go through head start, then KG before I felt that I was in first grade.

Those were the most difficult experiences of my life.

With GOD's help I feel I am doing well in my first grade.

God has granted me peace with myself, with GOD, and with GOD's Creation.

I just cannot wait to graduate and celebrate my first grade graduation celebration, and have a diploma signed by GOD.

If I treat God's creation the way God treats me, then God will grant me heavenly life in this world.

Healing Tip 354

* Special thanks to my teacher Nikitas J Zervanos, MD for being the inspiration to me to write this healing tip.

This world is a boot camp and offers different opportunities'.

Every human has a choice to be enrolled to be a spiritual being, human being type #1, human being type #2 or devil being.

Everyone is allowed to change enrolment to a different course.

To be enrolled in a spiritual being course the objective is to acquire vision of the heart and see the hidden perfection in everything and everyone.

Student spiritual people do good to those who do bad to them in order to earn credit toward graduation and live heavenly life.

Student devil being do bad things to good people to earn credit toward graduation and are unaware that their life is a living hell.

Healing Tip 355

Major danger to my wellbeing lies within me.

If I am not self-aware, then I am a grave danger to myself and others.

To become self-aware I must know my maker and follow his instructions to become self-aware, before I operate myself and stay connected and update myself to know any change and updates and practice those with best of my abilities.

It is not a manufacturing defect, but a safety feature to remind me to stay connected to the maker.

Healing Tip 356

I learned true forgiveness from my patient.

I saw this young lady for the first time in 25 years ago in a psych ward.

She shared her story with me, and it was mind boggling.

She grew up on a farm, as her father was a farmer.

She had a brother who was 4 years older than her.

Her mom and dad were never home, and her brother began molesting her.

He would perform oral sex on her and force her to reciprocate, and would always pay her as a reward for her good services.

He convinced me that nothing was wrong with it because Adam and Eve's kids did it

to each other.

He never thought this was bad, and this went on until age 20 when he moved out.

She became promiscuous, and began having sex for money.

A boy wanted to have sex with her, and she refused, and he told her she had no choice anymore because she was a prostitute.

She was shocked, and tried to commit suicide, and ended up in a psych ward.

I tried to explain that she could not punish herself for the wrong doing of others.

To me, she was innocent.

A couple years later she saw me in my office several times.

I encouraged her to go to college, and she did.

Last year she called me, and told me her success story.

She has master degree, and is doing very well.

She is married, and has three children.

Her brother was in a car accident, and was paralyzed from the neck down. She was his caregiver.

I asked how she was able to do what she did.

She replied: "I really owe my success to my brother. He put me through hell,

And I had a choice to stay in hell or use the raw material I received to build my own heaven.

I tried to kill myself and was unsuccessful, so I did not have a choice except to move on.

Now, I look at my brother lying helpless with tears rolling in his eyes.

He is in hell and he is reflecting, repenting, and trying to discover his heaven."

Healing Tip 357

Human life begins at the 4th week of gestation, commencing with the heart beat.

The difference between the human heart and animal heart is that human heart has a spiritual part as well as a physical one.

The Lord resides in the spiritual heart, and is the guarder of the soul.

The heart functions as unit without the brain, as brain function develops much later.

Both hearts need healthy nutrition to thrive, and this can be found through prayer.

Until the age of 25, surroundings bring a major impact on the physical and spiritual health of a person.

It is crucial to be aware of your spiritual heart.

Without this awareness, you may live your life and die without knowing the difference between the animal and human life.

However, humans can have the personality traits of animals.

Depending on the traits you have, you may be similar to a pig, cow, or even a bird.

With awareness, you can feed your spiritual heart and become a spiritual being by living a true life instead of just occupying space and serving a lower purpose than you are able to serve.

Healing Tip 358

Special thanks to a very special, beautiful young girl who I will call Jane Smith for confidentiality purposes.

She has been a great inspiration to me, thanks to the letter she wrote me on June 25th, 2015:

"Dr Shaikh,

Your book has taught me so much about unconditional love and forgiveness.

Your ability to see the good in everyone has left a lasting impression on me.

I've found myself troubled by wounds of the past, unable to move forward and forgive.

Your unwavering commitment to self growth and faith in people has helped me to release myself

from anger and resentment in order to move forward.

Through our visits and your teachings, I remember to see the good in people.

—Jane Smith."

She got more from that book than I did.

She is young with so much wisdom, and I wish I could be as wise as her.

What a beautiful letter, these words came through the clouds to provide hope,

love, and forgiveness.

If she can be this hopeful, what can stop any of us from doing the same?

Healing Tip 359

Open question to all Americans:

If today was the election day, and the candidates for presidency were Jesus of Lucifer, whom would you vote for and why?

To be qualified to vote for Jesus, one must practice his commandments.

Jesus received the popular vote by 98%, but his voters were only qualified by 1%.

Lucifer was only voted by 2% of the population, but received the presidency.

Healing Tip 360

I would like to thank two very special people who have inspired me to write this healing tip: Pat Botti and Shannon Quigley.

Once upon a time there was a righteous king.

There was a sign on the front of his castle that read:

"For righteous people this world is a prison,

For others it is a paradise"

One day, a prisoner was presented to him.

The prisoner said to the king: "I am not righteous and I am in prison. What kind of paradise is that? You are the king and live in a palace. What kind of prison is that?"

The king said: "My prison is a lot more difficult than yours. I wish to be in prison so I can worship and be with my creator 24/7.

But, the lord chose me to serve my time as a king, and so I must serve my time well so I can earn my blessings and be able to go back home to heaven instead of ending up in the hole of hell.

My prison is a lot more difficult than yours."

Then, he advised the prisoner: "The lord has given you an opportunity to reflect and serve your time well. This prison may turn into a heaven, and you will eventually go to heaven."

Anyone can take a U turn and become paradise bound. Not only is it legal, but it is highly recommended.

Healing Tip 361

With self awareness and self love, an unhealthy person can become healthy and live a happy life.

With self purification, a sinner can become a saint and live a peaceful life.

With sincere repentance, even Lucifer can become the chief of angels again.

Healing Tip 362

MYSTRY

MY

S - solution to solve mystery is self-purification and turn into spiritual being.

TRY – Stands for 'it's never too late to start trying self-purification and stay pure

Healing Tip 363

OPEN LETTER TO ALL HUMANS
* I AM THE LORD *
ALL POWERFUL – ALL WISE – ALL MERCIFUL

I created heavens, earth, sky, stars, planets, animals, Lucifer, and everything with perfect wisdom.

Everything and everyone is my creation.

I decided to make my masterpiece HUMANS and blow a tiny fraction of soul into them.

Every human has the potential to become a spiritual being – what I want from each one of them.

Humans are made of dirt, and the flesh made of dirt always inclines to be flesh driven—stomach, genitalia, and tongue are all flesh.

Unaware, humans live a flesh-driven life. Their lives are like a rat race—even if one wins, he's still a rat.

With awareness, humans know they are spiritual beings, living a heavenly life and serving a purpose.

THE HEAVENLY ENDING

From LORD (Signature on file)

To talk to the LORD directly, please call 1-800-Cre-ator. Use only the heart phone, and the heart must be pure.

PS: The biggest discovery to me is discovering that I can turn into a spiritual being by purifying my heart.

The Lord resides in a pure heart, and not only will mind and body change, but all aspects of life will change.

Addendum: Be a spiritual being. Live a heavenly life or a worldly life—each human being has this choice.

Healing Tip 364

From Lucifer the Great (Ex Chief saint) Ruler of the world Master of Most Humans.
TO: The Lord
CC: all Humans
One of my human slaves showed me the letter you wrote, and I cannot stop laughing. I agree that the Lord is all powerful and all merciful, but with all due respect, I disagree that he is wise.

It told him that humans would create bloodshed, commit sins, and cause suffering. The lord not only disagreed with me, but you asked me to bow down to humans.

I only bow down to you my lord.

I could not understand your wisdom then, and I still do not understand it now.

Innocent people are suffering, kids are dying, and many people are dying from diseases and sickness. It breaks my heart.

I told the lord this would happen.

Face it LORD, I am better than humans.

Better than you.

If I find one human who is better than me, then I will gladly bow down to them.

However, I'm sorry because I still haven't found that human yet.

I am assigned a dirty job: to lead people astray. It is the easiest job, as they perform it even better than me.

And I am considered the bad guy. Where is the justice? Where is the wisdom?

I know on the day of judgment, You have to apologize to me, but I do not need Your apology, as you are my Lord.

I do not want to be humiliated in front of everyone.

I am willing to forgive you, My Lord, if you have mercy of humans.

I want to be in your company again, I miss it so much.

Humans are nothing but trouble and the root of all evil.

Have mercy and put an end to this nonsense.

Your most faithful servant,
Lucifer

Healing Tip 365

From Arif M Shaikh M.D.

Dear Lucifer,

I read your letter, and as a physician, I see that you are sick and need a good doctor. I will treat you free because I have learned so much from you. I am not better than you. Only the Lord knows who is better, for the Lord is the only judge. You committed the original sin, disobeyed the Lord, and refused to bow down. Instead of repenting, which Is the universal antidote to all issues, you thought you were wiser than the Lord. That is not correct. I can feel your pain, but you have created your own pain, and have brought pain to all humans.

My recommendation is to try my prescription. If you donot get the desired effects, you will get a 100% money back guarantee, no questions asked.

Rx is sincere repentance.

You have nothing to lose, and it may be a win-win for everyone.

There was an inmate in prison who was an ex-criminal like you are, and now he's an angel as you were. You can learn from him and bow down to him, maybe you will become an angel the way you were and put an end to all the suffering that you are creating

Thank you Lucifer for showing me what is wrong with myself. I sure owe you a lot. I wish you all the best in whatever you choose to do.

Sincerely yours,
Arif M Shaikh M.D.

Dear Ex-Lucifer/Devil

I read your letter and I am impressed you feel you are superior to humans because you are master of most human.

Are you superior to me? It is true that your past is glorious lot better than my past. Your present is horrible, my present is not so horrible, so we are equal.

I am a spiritual being in a human body. If I am able to do self-purification, repentance with the Lord's help, I will become a spiritual being and my present will be the same as your past was.

If you do the same, you present will become the same as your past was and you may get your old job back.

I met many inmates who are better than you and me. Their past is worse than your present. Eviler than you but they are being judged and are blessed serving their time. I know a few inmates whose present is better than your past, if you learn from them, and bow down to human, you will become real lucifer and get your old job or wait until Lord will make the final decision. Most encounter with humans I feel they are ok and I can learn from them to be ok. In your case, I feel I am not neither are you. Lord knows the best wish and pray for the best hope you will repent.

Hope to see you in heaven.
Sincerely,
Your well wisher
Arif Shaikh

www.ingramcontent.com/pod-product-compliance
Lightning Source LLC
LaVergne TN
LVHW091545060526
838200LV00036B/707